EMERGING NANOTECHNOLOGIES IN FOOD SCIENCE

EMERGING NANOTECHNOLOGIES IN FOOD SCIENCE

Edited by

ROSA BUSQUETS
Kingston University London,
Kingston upon Thames,
United Kingdom

ELSEVIER

Elsevier
Radarweg 29, PO Box 211, 1000 AE Amsterdam, Netherlands
The Boulevard, Langford Lane, Kidlington, Oxford OX5 1GB, United Kingdom
50 Hampshire Street, 5th Floor, Cambridge, MA 02139, United States

Library of Congress Cataloging-in-Publication Data
A catalog record for this book is available from the Library of Congress

British Library Cataloguing-in-Publication Data
A catalogue record for this book is available from the British Library

ISBN: 978-0-323-42980-1

Publisher: Matthew Deans
Acquisition Editor: Simon Holt
Editorial Project Manager: Sabrina Webber
Production Project Manager: Julie-Ann Stansfield
Designer: Greg Harris

Typeset by Thomson Digital

To Pilar and Nicolau Ros

CONTENTS

LIST OF CONTRIBUTORS

Lorenzo Bautista
LEITAT Technological Centre, Terrassa, Spain

Ana I. Bourbon
University of Minho, Braga, Portugal

Diana M. Bowman
Arizona State University, Phoenix, AZ, United States

Rosa Busquets
Kingston University London, Kingston upon Thames, United Kingdom

Miguel Ângelo Cerqueira
International Iberian Nanotechnology Laboratory, Braga, Portugal

Frédéric Coulon
Cranfield University, Cranfield, United Kingdom

Andrew B. Cundy
University of Southampton, Southampton, United Kingdom

Samuel Eduok
Cranfield University, Cranfield, United Kingdom; University of Uyo, Uyo, Nigeria

José Manuel García
LEITAT Technological Centre, Terrassa; University of Barcelona, Barcelona, Spain

Joseph Lacey
Rayner Intraocular Lenses Limited, Worthing, United Kingdom

Jose López
LEITAT Technological Centre, Terrassa, Spain

Karinne Ludlow
Monash University, Clayton, VIC, Australia

Joseba Luna
ONYRIQ, Barcelona, Spain

Lubinda Mbundi
Blond McIndoe Research Foundation, Queen Victoria Hospital, East Grinstead, United Kingdom

Laura Molina
LEITAT Technological Centre, Terrassa, Spain

Sandra Niembro
LEITAT Technological Centre, Terrassa, Spain

Ana C. Pinheiro
University of Minho, Braga; Instituto de Biologia Experimental e Tecnologica, Oeiras, Portugal

Oscar L. Ramos
University of Minho, Braga, Portugal

Hélder Silva
Unilever UK Limited, Leatherhead, United Kingdom

Peter Viberg
Awapatent AB, Lund, Sweden

Antonio A. Vicente
University of Minho, Braga, Portugal

Alejandro Vílchez
LEITAT Technological Centre, Terrassa, Spain

PREFACE

It has been a great pleasure to work with other authors of this book, *Emerging Nanotechnology in Food Science*, which is designed to give a comprehensive understanding of the developments in nanotechnology relevant to food. The presence of nanotechnology in the food products is a newly emerging phenomenon and it is progressing at a fast pace as a result of increased appreciation of the advantages associated with the inclusion of nanomaterials in food, food contact materials, and food production processes. Therefore there is the need to collect and analyze the current knowledge to inform professionals coming from different sectors.

Nanotechnology has incidence in many facets of food science today and as such this book has taken an interdisciplinary approach so as to give the readers a broad understanding of food, chemical, biological, and legal aspects of nanotechnology in food and the interplay among them. This book splits this broad topic into chapters that deal with the latest nanotechnological innovations in food and food packaging materials, and the current knowledge on analysis, toxicity, regulatory framework, and intellectual property with rigor and depth. This is done by a team of food scientists, chemists, geochemists, biomedical scientists, biologists, engineers, and lawyers with recognized expertise in the areas of food and nanotechnology. The authors have sought to explain the complex concepts of nanotechnology in food in a didactic manner to make the content of the book attainable to a wide audience.

The first chapter covers basic concepts of nanotechnology (Chapter 1) and initiates readers who are not familiar with the subject of nanoscience and applications thereof, which will help with the comprehension of the rest of the book. Chapter 2 is devoted to the application of nanotechnology in food products. A separate chapter includes the advances in the production of safer water by means of nanotechnology solutions (Chapter 3). The gain of knowledge in any discipline greatly depends on the capacity that there is to measure data and acquire information. For that reason, a chapter (Chapter 4) has been dedicated to review the trends in the analysis of nanomaterials in food and to explain where the boundaries of such measurements are. One of the main aspects that is holding back putting in practice the advances of nanotechnology in food is the incomplete knowledge on the safety of products containing nanomaterials. For that reason, the fate of common nanomaterials in our body and toxicity at cellular level has been addressed in Chapter 5. Chapter 6 deals with the specific case of the toxic effect of nanomaterials on microbes, which can be used to our benefit through the manufacture of products incorporating antibacterial properties that increase food product shelf life and lead to a reduction of food waste. Increasing shelf life is a very important goal of food packaging, coatings and inks incorporating nanotechnology and have formed the main focus of Chapters 7 and 8. The manner in which nanotechnology is regulated and the principles underpinning it are discussed in Chapter 9. In addition, the book could not be complete without a chapter in intellectual property, which drives innovations generated in research laboratories to their exploitation and investment in new research that will make advance food nanotechnology. This is addressed in Chapter 10, which critically reviews relevant patents in the manufacturing and use of nanomaterials in food-related applications and analytical system, and discusses trends in intellectual property in the food sector.

The didactic approach followed in this book will facilitate the comprehension of rather specific and complex topics and make it accessible and suitable for either specialist or lay readers. The inclusion of the most recent research advances in the different facets of nanotechnology in food gives a unique contemporary understanding of this dynamic field.

CHAPTER ONE

Concepts of Nanotechnology

Rosa Busquets*, Lubinda Mbundi**
*Kingston University London, Kingston upon Thames, United Kingdom
**Blond McIndoe Research Foundation, Queen Victoria Hospital, East Grinstead, United Kingdom

Contents

1.1 INTRODUCTION

Nanoscience is the study of materials in the nanometer scale, which is generally considered to be a size below 1 micrometer (1 μm, a millionth of a meter). A nanometer (nm) is one billionth of a meter (10^{-9} m) and is the most commonly used size unit in nanotechnology. To put these units into context, a single standard piece of paper is approximately 75,000 nm thick, and if the earth and a double decker bus were scaled down a billion times, they would be the size of an olive and a nanoparticle, respectively (Figs. 1.1 and 1.2). Nanotechnology involves the production, manipulation, use, and characterization of nanomaterials. In this regard, a nanoparticle is defined as any material with at least one dimension measuring 1–100 nm and a nanomaterial refers to materials that contain nanoparticles and/or distinct features with at least one dimension measuring between 1 and 100 nm [1–3].

Although size is important in defining nanoparticles, when the particles reduce beyond a certain size, they inherit (quantum mechanical) properties that are different from those of bulk materials or atoms thereof. This is well accepted to be an important characteristic of nanoparticles at the size range between 1 and 100 nm, which has been adopted in the definition of nanoparticles and nanomaterials to reflect the importance of these properties. For instance, gold in bulk form is considered a *noble* metal due to its high stability and lack of reactivity but has different electronic properties and reactivity at nanoscale size [4,5]. It is these nanoscale-related properties that give these minuscule materials their *gigantic* potential in various applications.

Although nanotechnology is relatively new discipline that developed with the advent of microscopy and continues to be advanced further, nanoparticles have always

Figure 1.1 ***Comparison between the diameter of two systems with a billionth difference between their diameters.***

Figure 1.2 ***Comparison between the length of a double decker bus and a nanoparticle (iron oxide nanoparticle obtained with scanning electron microscopy).***

been there and continue to exist in nature. Naturally, minerals and crystals grow from the atomic to the macroscopic level passing through a stage where the crystals are in the nanoscale range and these can be found in the environment such as in volcanic ash or ocean spray. In space, NASA detected the presence of the carbon nanoparticles fullerenes [6], and possibly graphene [7]. Besides, anthropogenic activities continue to produce incidental formation of nanomaterials such as the spontaneous generation of nanoparticles from man-made objects. For instance, silver nanoparticles can emerge from objects such as sterling silver cutlery or jewellery [8], but today, one of the main sources of incidental nanomaterials may come from particulates emitted from running diesel engines [9].

Nanoparticles have also been found in human art and products manufactured before the advent of nanoscience. Notable examples of early manufactured objects with nanotechnology are the Lycurgus cup, from the times of the Roman Empire, known for its characteristic color changes when shone with light from different directions [10]. This cup, which is on display at The British Museum (London), has an opaque greenish-yellow tone when reflecting direct light, and it gives a translucent ruby color when light passes through it. This is caused by the presence of traces of submicroscopic crystals of silver and gold (so-called *colloids*) present at a gold to silver ratio of 3:7, specifically. Colloids are particles with at least one direction with dimensions comprised between 1 nm and 1 μm. Although silver nanoparticles are responsible for the greenish tone, the translucent reddish is attributed to the gold nanoparticles. These nanoparticles, along with the other traces of copper, antimonium, and sodium chloride nanoparticles, were not added deliberately but formed during the conditions employed when manufacturing the cup because nanotechnology did not exist then.

Another activity that, in a way, recognized the properties resulting from nanomaterials was the manufacture of glass as early as in Medieval Age: nanogold has been found in medieval stained glass, and it has been suggested to have played a role in purifying air through photocatalysis given that nano gold can break down volatile organic compounds [11]. A similar case explains the presence of nanomaterials in steel blades used in the Crusades [12]. Such nanomaterials are thought to be responsible for the exceptional mechanical properties of that steel. Damascus steel, renowned for its strength, has been found to contain carbon nanotubes.

These incidental and natural nanoparticles present a broad range of compositions, shapes, and sizes, which contrasts with these which have been engineered. Advances in the preparation and characterization of nanoparticles have allowed their engineering or optimization toward a goal that is to achieve desired properties and made possible their controlled production [2]. Such properties may depend on controlled charge, narrow size distribution, and a determined structure and composition of the particles at the nanoscale. These properties will translate into superior performance at the macroscale. For instance, the optimization of nanoparticles to achieve optimal size and reactivity allows having highly effective water filter or barrier properties in food packaging, as it will be discussed in Chapters 3,7, and 8.

The advancements of techniques such as microscopy [i.e., confocal, scanning electron microscopy (SEM), transmission electron microscopy (TEM)], dynamic light scattering, and elemental analysis have led to the increased interest in nanomaterials within and across different disciplines such as physics, chemistry, biology, engineering and arts, to name but a few, where they are studied and used in a wide range of applications. In this regard, nanomaterials have seen increased use in the latest advances in food, textile, health, electronics, energy, and environment. The application of nanotechnology continues to evolve with the inevitable advancement of nanoscience and public interest as more economic opportunities are realized. Given the multidisciplinary nature of nanotechnology, the vastness of the discipline in both fundamental development and application, and the need to communicate the technology to lay users, there has been increased calls for the development of basic and clear nanotechnology concepts to aid dissemination and regulation.

According to the current definition of a nanomaterials, all materials containing nanoparticles in unbound, aggregate, or agglomerate states, which impart key properties, and those that have structures (i.e., internal or external pores or fibers) in the nanoscale range can also be considered to be nanomaterials [1–3]. As a result, countless objects can be considered to be nanomaterials given that many materials contain several distinct nanosized structures. Although this broad definition highlights the ubiquitous nature of the nanomaterials, it also creates regulatory difficulties of producing clear guidelines that both encompass all aspects of nanomaterials as per definition and are clearer. According to European Chemical Industry Council, the accepted definition for nanomaterials is still too broad in scope [13]. The FDA has not published regulatory definitions [14] and is less specific in their approach than the ones by the European Commission [3]. As such, an attempt to narrow the definition, a suggestion that a material should have at least 50% of the particles it contains measuring at least 100 nm in one dimension in order for it to be described as a nanomaterial is under review [1–3]. While the argument for the 100 nm upper limit in the definition of nanoparticles is to do with special and unique properties below this size, the argument for not including particles with one or more dimensions below 1 nm remains unclear and contentious. Indeed, subnanometer particles, such as fullerenes, with unique properties resulting from their size and structure are well established and have been recommended for classification as nanomaterials [3]. This is a clear demonstration of the dynamic nature nanotechnology and the discipline evolves. However, not every item with dimension below 100 nm can be considered a nanoparticle. Currently, submicron molecules such as proteins and cellular organelles, although below 100 nm, are not regarded as nanomaterials as they do not have *special* physico-chemical properties of nanoscale materials (i.e band gap) [15].

1.2 SIZE MATTERS

The cause of the special properties of nanomaterials lies in the dimensions of nanomaterials: they are much smaller than bulk materials and bigger than individual atoms. As a result their physico-chemical behavior do not follow classical physics (which describes properties in the macroscale) or quantum chemistry (which describes phenomena at atomic level). A direct consequence of the reduced size of nanoparticles is greater surface-area-to-volume ratio than macroscopic materials, which is characterized by higher number exposed atoms that can participate in chemical reactions. In Fig. 1.3, the surface-area-to-volume ratio of a three-dimension cubic microparticle (1000 nm per side) (Fig. 1.3A) is compared to the surface-area-to-volume ratio of the 1000 loose nanoparticle building blocks (100 nm per side) (Fig. 1.3B) that form the microparticle and its mass. In this case, the combined surface-area-to-volume ratio of the loose nanoparticles is 10 times greater than that of the microparticle 1000 nm^3 cube. This property is economically attractive as it allows a given mass of nanosized particles to be several times more reactive and efficient than the equivalent bulk mass. In addition, defects in their structure of a nanoparticle and the resulting loss of stability can also increase reactivity. Hence, atoms which have lost some of its neighboring atoms, and therefore have incomplete coordination, can result

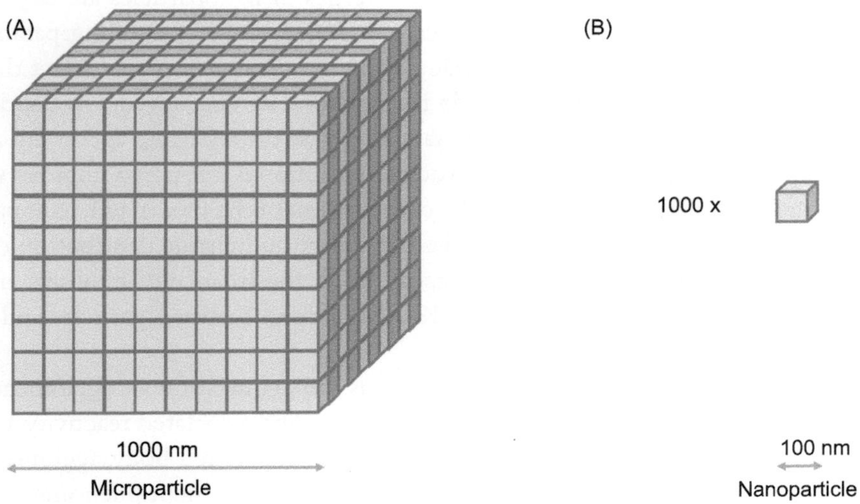

Figure 1.3 ***Comparison between the surface area of a microparticle and a nanoparticle.*** The microparticle (A), with 1000 nm side, has a total surface area of 6,000,000 nm^2. The microparticle can be broken down into its 1000 nanoparticle components (B). The surface area of each nanoparticle (B), with 100 nm side, is 60,000 nm^2. Therefore, the disaggregated group of nanoparticles integrating the microcube can provide a surface area of 60,000,000 nm^2.

in defects in the material such as dislocations in the crystal structure, or result in the inclusion of impurities. This can introduce new edges in the crystal and or impart new properties that can enhance reactivity of the whole material. However, the mechanisms underlying these effects and how to control them are still unclear [16].

Nonetheless, these unique properties continue to be exploited for different application. Good examples of this are nanosized metals and semiconductors whose electrons and holes (also called charge carriers) are confined in a limited space, which causes the splitting of the edge of the valence and conduction band into quantised electronic levels. The valence conduction band plays an important role in the electric and optic properties and the quantised electronic state creates an intermediate state between bulk materials and atoms or molecules. The spacing between these electronic levels and band gap (between the valence and conduction band) increases with a decrease in particle size. As a consequence, more energy is required to transfer electrons from the valence band to the conduction band in a nanoparticle than in a microparticle. An example of the effect of particle size on the electronic properties is the different color that some nanoparticles present at different sizes. For instance, although gold in the bulk scale presents golden color, nanoparticles of gold present a range of colors between red and pink, depending on their size and how they interact with light and electron clouds of atoms [5]. Given that light absorbed is affected by the size of some nanoparticles, color changes can be used to determine changes in particle size and certain particle properties related to the structure. Moreover, given that several other properties of nanoparticles are dependent on size and stability, desired characteristics can be imparted by controlling parameters that affect these features during nanoparticle production. Parameters that affect the size of nanoparticles during production include temperature, stirring speed, type, and concentration of reducing agent as well as the rate at which the reducing agents are added [17,18]. In addition, the stability of nanoparticles in solution is known to improve with a reduction in particle size and more so by the composition of the media, pH, nanoparticle concentration, and interparticle distance. The net charge and charge distribution on the surface of particles, which is pH dependent, can affect the balance of repulsive and attractive forces (i.e., Born, Van der Waals, or Keesom forces) between particles and favor their precipitation or suspension in solution.

The dispersion of nanoparticles is important as the aggregation or agglomearation of particles can reduce the surface area of nanoparticles and associated reactivity. Unlike aggregation that leads to formation of new particles with sintered body, agglomeration (reversible weak physical adhesion of nanoparticles) [19] is a transient phenomenon that remains a challenge to characterize and control. Some of the common ways of controlling agglomeration include the use of ultrasonication shortly before the use of nanoparticles that are suspended in solution. Alternatively, the addition of surfactants during or after the preparation of the nanoparticles is also used to minimize agglomeration. In this case, a critical concentration of surfactant for optimum coverage of the nanoparticle surface is necessary to achieve better stability. Commonly used surfactants can be anionic [i.e., sodium dodecyl sulfate (SDS) or sodium citrate]; cationic (i.e., cetylpiridinium bromide); zwitterionc (i.e., lecithin), or nonionic (i.e., tetraoxyethylene lauryl ether).

1.3 HEALTH CONCERNS AND PUBLIC UNDERSTANDING

The toxicity of nanoparticles is linked with their unique properties which are dictated by their composition and structure (shape, electronic properties, particle size distribution, agglomeration, surface chemistry, concentration, stability) as well as the route and nature of exposure (temporal, intermittent, continuous exposure) [20–25].

The toxicity and poorly understood fate of nanoparticles in biological systems are some of the main factors that limit appreciation and use of nanotechnology. The public's poor understanding of nanotechnology and preconceived fears is another factor that can affect the speed and scope of the application nanotechnology. This in part because scientific advancements happen at a faster rate with new concepts emerging before the public have acquired enough understanding of the meaning and implications of the new technology. Genetic technology (i.e., cloning and genetically modified food) is one example of such scientific advance where the poor understanding and appreciation by the general public has resulted in strong opposition to the study and application of the technology [26]. On the other hand, the use of food additives, natural or synthetic, that are currently widely used to enhance flavor or color, shelf life and taste, are well received by the public, albeit the little understanding and anxiety. This gives some hope that nanotechnology could potentially be as well accepted, despite the anxiety associated with artificial food additives. A recent study pointed out that knowledge of regulation and trust in regulators, awareness of risks and benefits and preference for natural products influences the acceptance of artificial food additives by consumers [27]. This situation is relevant to the introduction and use of particular matter in the nanodimensions in food and related technologies and therefore advances in regulation and communication with the public will foster benefiting from safe food products incorporating nanotechnology.

The scientific community has the responsibility of communicating scientific advances and concepts such as nanotechnology and associated benefits and risks of its incorporation in food products. This can minimize the confusion and anxieties that cause the public to not only dislike but also resist scientific advances such as nanotechnology. The public needs to know the whole picture before accepting a technology as potentially beneficial and reliable. Indeed, if the new scientific concept remains misunderstood, related technologies will advance with difficulties, if not advance at all. Nanotechnology is an emerging technology with immense potential to advance a number of current technologies, including food science, and should be embraced and explored thoroughly.

REFERENCES

[1] European Commission, Definition of a nanomaterial, 2015. Available from: http://ec.europa.eu/environment/chemicals/nanotech/faq/definition_en.htm

[2] ISO, Nanotechnologies—Vocabularies for science, technology and innovation indicators, ISO/TS 18110:2015(en), 2015. Available from: https://www.iso.org/obp/ui/#iso:std:iso:ts:18110:ed-1:v1:en:term:2.8

[3] The European Commission, Commission recommendation of 18 October 2011 on the definition of nanomaterial, Off. J. Eur. Union (2011) 2011/696/EU, L275/38-L275/40.

[4] H. Hu, L. Reven, A.D. Rey, DFT Study of gold surfaces–ligand interactions: alkanethiols versus halides, J. Phys. Chem. C. 119 (2015) 11909–11913.

[5] A. Jimenez-Ruiz, P. Perez-Tejeda, E. Grueso, P.M. Castillo, R. Prado-Gotor, Non functionalized gold nanoparticles: synthetic routes and synthesis condition dependence, Chem. Eur. J. 21 (2015) 9596–9609.

[6] J. Cami, J. Bernard-Salas, E. Peeters, S.E. Malek, Detection of C60 and C70 in a young planetary nebula, Science 329 (2010) 1180–1182.

[7] NASA., C. Jet Propulsion Laboratory, Pasadena, Honeycomb carbon crystals possibly detected in space, Spitzer. Stud. Universe Infrared. Mission News (2011). Available from: http://www.nasa.gov/mission_pages/spitzer/news/spitzer20110815.html

[8] R.D. Glover, J.M. Miller, J.E. Hutchison, Generation of metal nanoparticles from silver and copper objects: nanoparticle dynamics on surfaces and potential sources of nanoparticles in the environment, ACS Nano 5 (2011) 8950–8957.

[9] A. Liati, S.S. Pandurangi, K. Boulouchos, D. Schreiber, Y. Arroyo Rojas Dasilva, Metal nanoparticles in diesel exhaust derived by in-cylinder melting of detached engine fragments, Atmos. Environ. 101 (2015) 34–40.

[10] I. Freestone, N. Meeks, M. Sax, C. Higgitt, The Lycurgus cup—a roman nanotechnology, Gold Bull. 40 (2007) 270–277.

[11] X. Chen, H.Y. Zhu, J.C. Zhao, Z.F. Zheng, X.P. Gao, Visible-light-driven oxidation of organic contaminants in air with gold nanoparticle catalysts on oxide supports, Angew. Chem.-Int. Ed. 47 (2008) 5353–5356.

[12] M. Reibold, P. Paufler, A.A. Levin, W. Kochmann, N. Pätzke, D.C. Meyer, Materials, Carbon nanotubes in an ancient Damascus sabre, Nature 444 (2006) 286.

[13] Organisation for Economic Co-operation and Development. OECD, Environment, Health and directorate. Joint meeting of the chemicals committee and the working party on chemicals, pesticides and biotechnology. Developments on the safety of nanomaterials 2013, n 59, ENV/JM/MONO (2015) 41, Paris, 2015, pp.1–86.

[14] Food and Drug AdministrationGuidance for industry considering whether an FDA-regulated product involves the application of nanotechnology, Biotechnol. Law Rep. 30 (2014) 613–616.

[15] Executive Office of the President of the United StatesNational Science and Technology Council Committee on Technology Subcommittee on Nanoscale Science, Engineering and Technology. National nanotechnology initiative strategic plan, Natl. Nanotechnol. Initiat. (2014) 1–88 http://www.nano.gov/sites/default/files/pub_resource/2014_nni_strategic_plan.pdf

[16] W. Yu, J. Zhang, T. Peng, New insight into the enhanced photocatalytic activity of N-, C- and S-doped ZnO photocatalysts, Appl. Catal. B Environ. 181 (2016) 220–227.

[17] M. Hayat (Ed.), Colloidal gold, Principles Methods and Applications, vol. 1, Academic Press Inc, San Diego, CA, London, 1989.

[18] D. Li, R.B. Kaner, Shape and aggregation control of nanoparticles: not shaken, not stirred, J. Am. Chem. Soc. 128 (2006) 968–975.

[19] Commission for the Investigation of Health Hazards of Chemical Compounds in the Work Area Report, Nanomaterials, Wiley-VCH GmbH& Co. KGaA, Bonn, Germany, 2013.

[20] G. Oberdörster, Toxicology of ultrafine particles: in vivo studies, Phil. Trans. R. Soc. Lond. 358 (2000) 2719–2740.

[21] C.M. Sayes, F. Liang, J.L. Hudson, J. Mendez, W. Guo, J.M. Beach, V.C. Moore, C.D. Doyle, J.L. West, W.E. Billups, K.D. Ausman, V.L. Colvin, Functionalization density dependence of single-walled carbon nanotubes cytotoxicity in vitro, Toxicol. Lett. 161 (1775) (2006) 135–142.

[22] C.M. Sayes, Correlating nanoscale titania structure with toxicity: a cytotoxicity and inflammatory response study with human dermal fibroblasts and human lung epithelial cells, Toxicol. Sci. 92 (2006) 174–185.
[23] P. Wick, P. Manser, L.K. Limbach, U. Dettlaff-Weglikowska, F. Krumeich, S. Roth, W.J. Stark, A. Bruinink, The degree and kind of agglomeration affect carbon nanotube cytotoxicity, Toxicol. Lett. 168 (2007) 121–131.
[24] P.N. Breysse, R.J. Delfino, F. Dominici, A.C.P. Elder, M.W. Frampton, J.R. Froines, A.S. Geyh, J.J. Godleski, D.R. Gold, P.K. Hopke, P. Koutrakis, N. Li, G. Oberdörster, K.E. Pinkerton, J.M. Samet, M.J. Utell, A.S. Wexler, US EPA particulate matter research centers: summary of research results for 2005–2011, Air Qual. Atmos. Heal. 6 (2013) 333–355.
[25] P. Pedata, T. Stoeger, R. Zimmermann, A. Peters, G. Oberdörster, A. D'Anna, Are we forgetting the smallest, sub 10 nm combustion generated particles?, Part. Fibre Toxicol. 12 (2015) 34.
[26] P. Williams, E. Stirling, N. Keynes, Food fears: a national survey on the attitudes of Australian adults about the safety and quality of food, Asia Pac. J. Clin. Nutr. 13 (2004) 32–39.
[27] A. Bearth, M.-E. Cousin, M. Siegrist, The consumer's perception of artificial food additives: influences on acceptance, risk and benefit perceptions, Food Qual. Prefer. 38 (2014) 14–23.

CHAPTER TWO

Advances in Food Nanotechnology

Miguel Ângelo Cerqueira, Ana C. Pinheiro*,‡, Oscar L. Ramos*, Hélder Silva†, Ana I. Bourbon*, Antonio A. Vicente***

*University of Minho, Braga, Portugal
**International Iberian Nanotechnology Laboratory, Braga, Portugal
†Unilever UK Limited, Leatherhead, United Kingdom
‡Instituto de Biologia Experimental e Tecnologica, Oeiras, Portugal

Contents

2.1 OVERVIEW

Nanotechnology offers a vast number of opportunities for the agri-food industry. The primary production section would benefit of new pesticides with improved action thanks to the use of nanotechnology; animal feeds with enhanced efficacy and higher nutrition value through the use of enriched supplements; smart sensors for the diagnosis of animal disease or for the detection of pathogenics in water. The food production sector could also benefit from improvements in the detection and control of chemical and microbiological hazards, thus promoting food safety while eventually leading to an improved market value of the foods.

Food packaging, one of the utmost operational areas using nanotechnology in the food industry, can now incorporate antimicrobials to extend shelf life through the development of active and intelligent packaging systems. Additional to the development of new contact materials being investigated to preserve food organoleptic properties, new packaging systems are being designed to act as sensors that will inform about the quality or freshness of the sealed food products [1,2].

The nutritional value of food products can be improved through the use of nanotechnology, where the incorporation of bioactive compounds (such as antioxidants, vitamins, polyphenols, omega-3 fatty acids, among others) in nanosized structures has shown to increase their bioavailability. This opens up the possibility of establishing personalized nutrition schemes with on-demand health requirements and allowing the consumers to safely choose food products based on their best interest. This is a clear trend today: there have been increasing number of publications in this area during the past 10 years (Fig. 2.1A).

2.2 NANOTECHNOLOGY FOR ENCAPSULATION

Some biologically active compounds, such as lipids, vitamins, peptides, fatty acids, antioxidants, minerals, and living cells such as probiotics, are available on the market in various forms claiming to provide health benefits in the prevention and treatment of a range of diseases. However, the administration of these compounds in their free form often results in a low systemic bioavailability which is mainly attributable to insufficient gastric residence time, low permeability, and/or solubility within the gut, as well as instability under conditions encountered in food processing (i.e., temperature, oxygen, or presence of light) or in the gastrointestinal (GI) tract (i.e., pH and presence of enzymes and other agents)—all of which limit its activity and potential health benefits of bioactive molecules [3,4]. The development of nanoencapsulation technologies offers possible solutions to improve the bioavailability of many functional compounds, by enhancing their absorption in the GI tract, which is critical for their effectiveness [5]. Encapsulation consists in the entrapment of a substance (i.e., bioactive molecule) within a carrier material [6]. The delivery of an active agent can be improved by means of its encapsulation, for instance by enabling the slow down of the degradation processes; protecting its functionality; or permitting targeted release in sites where absorption is desired [4,7]. Moreover, nanoencapsulation can provide some important advantages. The stability of labile bioactive substances (i.e., vitamins) can be enhanced by means of increasing their solubility through nanoencapsulation, which implies solubilizing a hydrophilic compound in hydrophobic matrices or vice versa. Nanoencapsulation can also act as a barrier between bioactive molecules with limited stability and the environment, thus protecting

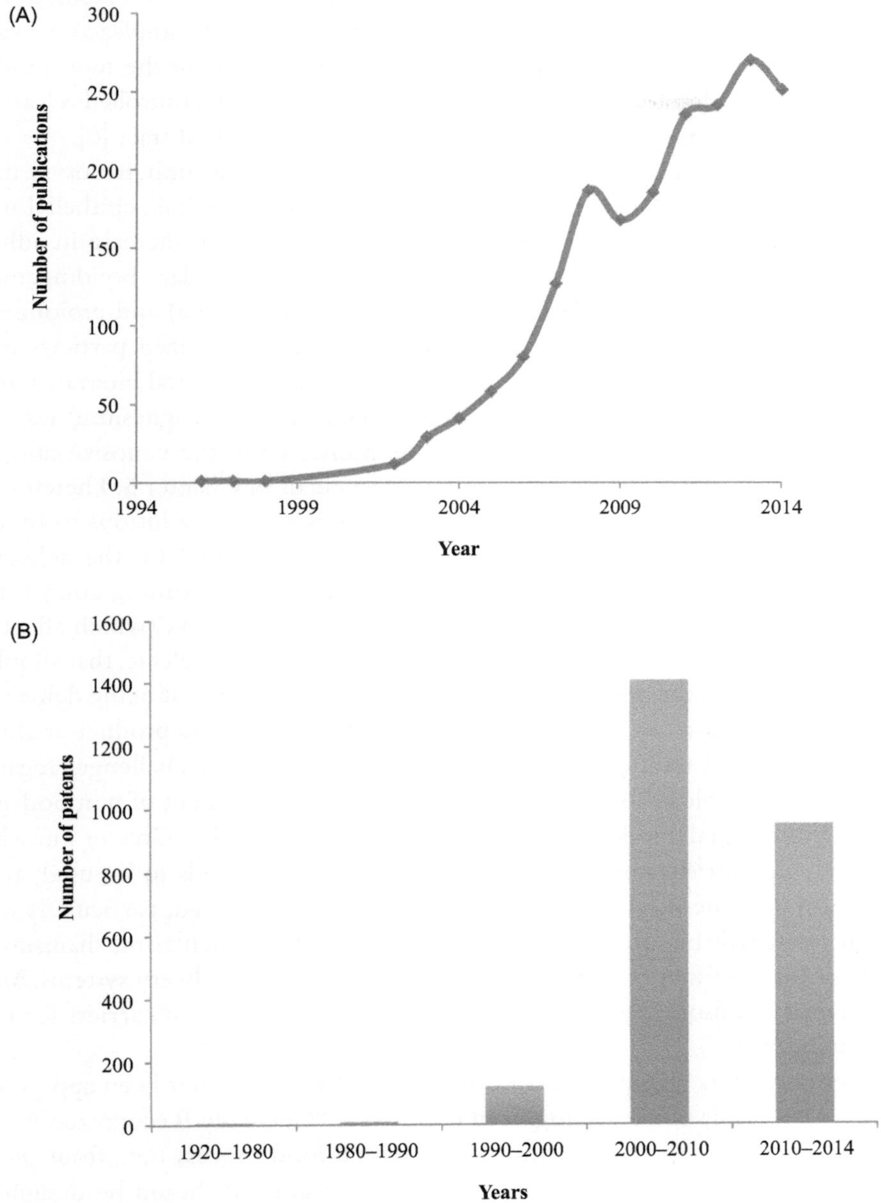

Figure 2.1 Number of publications (A) and issued patents (B). The keywords used were the combination of "food" and "nanotechnology." The source of information was Scopus database (keywords searched in the title, abstract, and keywords) and Google patents database (keywords searched in all fields).

against undesired permeation of gases like CO_2, O_2, H_2O, reducing the evaporation and degradation of volatile substances (i.e., aromas), as well as prevent unpleasant sensorial properties (i.e., bitter taste, astringency, or smell) when consuming the food products. Furthermore, this technology can be used for improved delivery, controlled release, and bioavailability of bioactive compounds to cells and tissues in the GI tract [6].

Due to their subcellular size and greater surface area per mass unit, nanosystems are expected to be more efficient in crossing permeability barriers (i.e., epithelial membrane cells and the tight junctions between them) including an increase in adhesive forces (i.e., interaction through physical entanglement and secondary bonding, mainly hydrogen-bonding and van der Waals attraction to the GI mucosa) and prolonged GI transit time, and present higher biological activity than larger-sized particles of the same chemical composition [8]. For instance, the absorption and oral bioavailability of several essential minerals (i.e., iron, zinc, calcium, selenium, and magnesium) was demonstrated to be significantly enhanced when administered in the nanosize range [9]. The bioavailability of nanomaterials is explained in detail in Chapter 5. Therefore, the development of nanoencapsulation technologies offers possible solutions to inherent difficulties associated with macro- and micro-scale encapsulation for the delivery of bioactive compounds. For instance, this technology allows overcoming compatibility issues (i.e., aggregation and phase separation) with the food matrix (which affects appearance, texture, stability, or flavor of the product), and control release, that should be only activated once at the appropriate site (i.e., compounds aimed at being delivered in the intestine should start its release as soon as mixed with the food product or affected by light, oxygen, and temperature) [5,10]. At present the major challenges regarding the use of this technology for food applications are the replacement of nonfood-grade materials by bio-based, biodegradable food-grade alternatives [8]. Greater knowledge of intrinsic characteristics of polymers and bioactive compounds to be used, as well as of physical and chemical interactions that could be established, particularly at the molecular level, will be crucial for understanding the fundamental mechanisms that are the basis for the functional properties and behavior of the delivery systems. All this is fundamental to ensure the design and development of optimized carriers for use in the food industry.

Different products widely used for drug encapsulation have not been approved for use in food as a result of not having been certified as "Generally Recognized As Safe" (GRAS) materials for food applications. The whole food process (i.e., from the raw materials until the final product that reaches the consumers) should be designed to meet the safety requirements of governmental agencies, that is, European Food Safety Authority (EFSA) or Food and Drug Administration (FDA) [11,12]. Products used for constituting the nanosystems must be bio-based, biodegradable, food-grade, and able to form a barrier between the internal phase and its surroundings. Polysaccharides (i.e., starch and their derivates, alginate, pectin, dextran, and chitosan) are the most widely used products. Proteins (i.e., gelatin, whey proteins, and zein) and lipids (i.e., medium chain triglycerides, tristearin, and corn oil) are also appropriate for encapsulation [6,8,13]. The features of encapsulating materials influence important characteristics, such as: physico-chemical properties (i.e., density, refractive index, rheology polarity, particle size, charge, and interactions); protective capacity (i.e., antioxidant activity); encapsulation performance (i.e., loading capacity and encapsulation and retention efficiencies); and properties related to the release of the bioactive substances (i.e., trigger, rate, and extent). In addition, the overall cost, safety, and composition of the encapsulating material should be taken into account with view of the final application [6,12,14].

There have been major advances in the design and production of food-grade nanosystems that can be used by food manufacturers to develop effective delivery systems [8,15]. Many encapsulation procedures have been proposed but none of them can be considered universally applicable for bioactive compounds. This is related to the fact that bioactive compounds and encapsulating material in the nanosystems have their own characteristic molecular structure and properties, which include molecular weight, polarity, solubility, particle size distribution, encapsulation efficiency, or shape. These properties can, in fact, be affected by the encapsulation techniques employed. Therefore, different encapsulation approaches may be required to achieve the desired physico-chemical and molecular properties [12,16].

The encapsulation techniques most commonly used are spray drying, spray cooling, freeze drying, emulsification, emulsification-solvent evaporation, coacervation, nanoprecipitation, inclusion complexation, and supercritical fluid [6,13,17,18]. These techniques use either top-down or bottom-up approaches for the development of nanomaterials. Top-down approaches involve the application of precise tools, assisted with a force, that allow size reduction and shaping of the structure for the desired application of the nanomaterial. The degree of control and refinement achieved in size reduction processes influences the properties of the materials produced. In the bottom-up approach, materials are constructed by self-assembly and self-organization of molecules, which are influenced by many factors, including pH, temperature, concentration, and ionic strength [19].

Techniques such as emulsification and emulsification-solvent evaporation are used in the top-down approach. On the other hand, the use of supercritical fluids, inclusion complexation, coacervation, and nanoprecipitation are used in the bottom-up approach [20]. Spray drying is by large one of the most frequently applied techniques because it is fast, relatively cheap, and reproducible. The principle of this approach is based on dissolving or dispersing the bioactive compound in a solution of biopolymer; the solution/dispersion is then atomized in a heated air chamber, which causes rapid evaporation of the solvent and production of dried systems. This method has some limitations for volatile or thermo-sensitive bioactive species [21,22]. Spray cooling may allow overcoming some limitations of spray drying as it involves an opposite principle. In spray cooling, a bioactive substance dispersed in a liquified matrix is atomized into a cool environment, such as cool air. Usually, fats with high melting point are used as matrix. When cooling, the fat solidifies, permitting the immobilization of labile compounds, such as mineral salts, enzymes, flavors, food acids, and protein hydrolysates [16,23]. Freeze drying is also an alternative to spray drying but usually restricted to very high value ingredients, such as probiotic bacteria. This technique involves the encapsulation of bioactive substances in a solution of biopolymer though a freezing process, which enables its immobilization, followed by drying under vacuum, which causes the sublimation of ice [16,23].

Emulsification is defined as a process of dispersing one liquid (containing the bioactive compounds) in a second immiscible liquid, by applying electrostatic, or hydrophobic, or hydrogen bonding interactions between the bioactive compounds and an encapsulating material [16]. The addition of a surfactant is also frequently used to promote encapsulation by forming micelles, vesicles, bilayers, and reverse micelles around the bioactive molecules [7,13,16].

The emulsification-solvent evaporation technique involves emulsification of the polymer solution into an aqueous phase and evaporation of the solvent. This induces the precipitation of the polymer in the form of nanospheres [24].

Coacervation is one of the most easily implemented techniques for the production of nanosystems consisting of electrostatic attraction between oppositely charged molecules. This force may be induced between charged bioactive compounds and an oppositely charged polymer (simple coacervation). Alternatively, a bioactive substance may be entrapped within a particle formed by electrostatic complexation of positively charged (i.e., chitosan) and negatively charged (i.e., pectin and alginate) biopolymers (complex coacervation). This technique is usually applied for the encapsulation of lipophilic flavors and oils and also some water-soluble bioactive compounds [3,4].

Nanoprecipitation involves the precipitation of polymer from an organic solution and the diffusion of the organic solvent in an aqueous medium. This technique is based on the spontaneous emulsification of the organic internal phase containing the dissolved bioactive compounds into the aqueous external phase [24].

Inclusion complexation consists of the encapsulation of the supramolecular association of a ligand (bioactive compound) into a cavity-bearing substrate (encapsulating material) through hydrogen bonding, van der Waals force, or by the entropy-driven hydrophobic effect [18].

In supercritical fluid processes for encapsulating bioactive substances, the bioactive compounds are dispersed in a matrix solubilized in a supercritical fluid (usually carbon dioxide under supercritical conditions). The removal of the carbon dioxide leads to the encapsulation of bioactive specie within the matrix [16].

Among the techniques described earlier, emulsification and emulsification-solvent evaporation are the most used techniques to promote size reduction since they have no impact on any other factor, such as pH and temperature. Coacervation, supercritical fluid, and inclusion complexation techniques also allow size reduction; however, these techniques are largely influenced by other factors (i.e., pH and temperature). Coacervation and supercritical fluid are used for encapsulation of both hydrophilic and lipophilic compounds, whereas emulsification-solvent evaporation and inclusion complexation are largely used for the lipophilic compounds [25].

Spray drying and freeze-drying are the most extensively applied encapsulation techniques in the food industry for the stable release of bioactive compounds. They have become essential because they are flexible, can operate in continuous mode, and are less expensive than other encapsulation techniques [6,26]. The drying process is characterized by providing greater stability to functional compounds compared to their stability in the original suspensions, which makes it possible to preserve their characteristics for long storage periods [27].

A selected group of bio-based nanodelivery systems used in innovative food products, which have been commonly used for encapsulation of bioactive compounds is reviewed in Table 2.1: nanocapsules, nanohydrogels nanoemulsions, lipid nanoparticles, and micelles. Fig. 2.2 shows some examples of nanosystems used for encapsulation.

2.2.1 Nanocapsules

Nanocapsules have been one of the most widely studied nanosystems for the delivery of functional compounds [43,44]. Nanocapsules (which are also known as nanoparticles in food science) are constituted by an external polymeric membrane and an inner part composed of a liquid or polymeric matrix that contains the bioactive compound [45]. During their production, different techniques can be applied (such as ionic pregelation/coacervation, polymerization, and dispersion of preformed polymers), which may affect their final properties [43].

Nanocapsules can be either synthesized through ionic pregelation/coacervation, which consist of crosslinking of polyelectrolytes in the presence of a counter ion (cationic or polyanionic) to form nanocapsules [8], or by the polymerization of monomers. In this latter technique, bioactive compounds can be incorporated in the capsules through their dissolution in the polymerization medium or by adsorption onto the nanocapsules after polymerization [24]. In addition, nanocapsules can be prepared by dispersion of preformed polymers and formation through different techniques, such as self-assembly, nanoprecipitation, and using supercritical fluids [16,46]. For instance, self-assembly involves the spontaneous formation of compact and stable nanocapsules without intervention from external agents. Biomaterials, such as zein, chitosan, and casein are examples of polymers that can be used in the preparation of nanocapsules via self-assembly mechanisms [8]. The preparation method for the development of nanocapsules must be optimized so that the final nanocapsules display higher performance [8]. Examples of nanocapsules prepared by different techniques using various bio-based materials are presented in Table 2.1.

Table 2.1 Examples of nanosystems for application in food products and their main characteristics: techniques and composition used, encapsulated bioactive compounds, and size

Nanosystem	Technique	Composition	Bioactive compound	Size (nm)	References
Nanocapsules	Ionic pregelation/ coacervation	Chitosan, alginate and sodium tripolyphosphate	Insulin	270	[28,29]
	Ionotropic polyelectrolyte pregelation	Alginate/ chitosan/pluronic	Curcumin	100–120	[30]
	Ionotropic polyelectrolyte pregelation	Alginate/chitosan	Vitamin B_2	86–200	[31]
Nanohydrogels	Physical self-assembly	β-lactoglobulin and low methoxyl pectin	ω-3 fatty acids	100	[32]
	Temperature and pH induced gelation	β-lactoglobulin	Epigallocate-chin-3-gallate	7–10	[33]
	Temperature induced gelation	β-lactoglobulin/ hen egg white protein/alginate	α-Tocopherol	–	[34]
	Temperature induced gelation	β-Lactoglobulin	Curcumin	142	[35]
Nanoemulsions	High-pressure homogenization	Corn oil, Tween 20, SDS, and DTAB	Curcumin	119.5–152.9	[36]
	Solvent displacement + ultraturrax	Hexane, Tween 20	ß-carotene	9–280	[37]
	Melt-homogenization	Medium chain triglycerides, trimyristin and tristearin	Curcumin	130–205	[38]
Solid lipid nanoparticles	Microemulsification	Polysorbate 80/ soy lecithin	Curcumin	134.6	[39]
	High-pressure homogenization	Soy lecithin/ sodium glycocholate/ glycerol	Curcumin	100–110	[40]
Micelles	Self-assembly	Casein	β-Carotene	80	[41]
		β-casein	Curcumin	—	[42]

Note: SDS: sodium dodecyl sulfate; DTAB:dodecyltrimethylammonium bromide; - Not found.

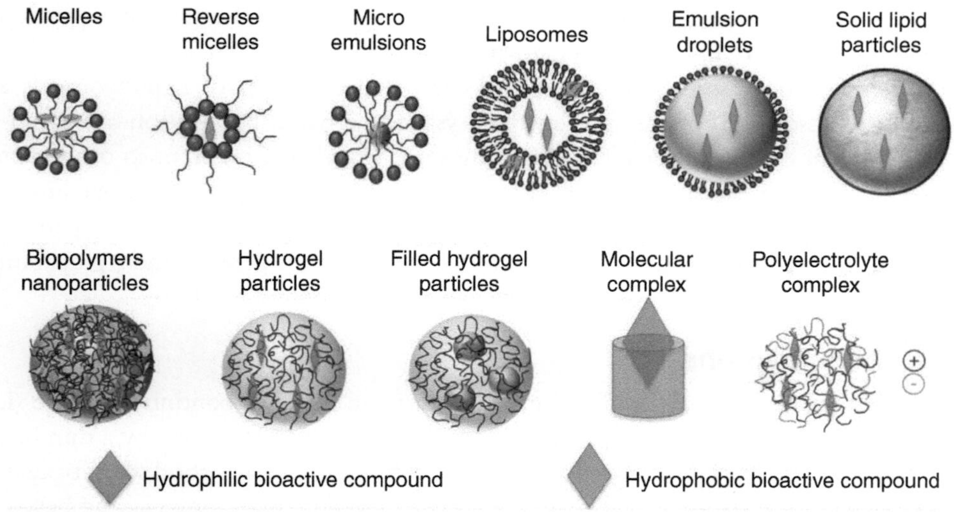

Figure 2.2 ***Examples of colloidal delivery systems that can be used to encapsulate, protect, and deliver functional food ingredients with the most likely location of the bioactive compound based on its hydrophilicity or hydrophobicity (not drawn to scale).*** *(Reproduced from I.J. Joye, G. Davidov-Pardo, D.J. McClements, Nanotechnology for increased micronutrient bioavailability, Trends Food Sci. Technol. 40 (2) (2014) 168–182 [14] with permission from Elsevier).*

2.2.2 Nanohydrogels

Nanohydrogels are three-dimensional hydrophilic or amphiphilic polymer networks that can swell in water (around 30 times their size), and hold relatively large amount of aqueous solvent, while maintaining their structure, due to the presence of covalent or noncovalent interactions [47]. Their swelling ability is attributed to the presence of hydrophilic moieties (i.e., hydroxyl, carboxyl, ethers, amines, and sulfate groups) forming the nanohydrogel structure, which is responsible for their soft and pliable attributes. Nanohydrogels are designed to spontaneously load biologically active molecules through electrostatic, van der Waals, and/or hydrophobic interactions between the bioactive compound and the matrix during the gel formation. This will result in stable nanostructures in which bioactive compounds become entrapped. These nanosystems can be prepared from several polysaccharides, such as alginate, chitosan, pectin, pullulan, and dextran or proteins, such as whey proteins and collagen, with different techniques.

Nanohydrogels have the ability to produce a response (i.e., swelling or deswelling) to environmental stimuli (i.e., temperature, pH, ionic strength, or enzymatic conditions) making them important systems to deliver bioactive food ingredients to specific sites of action and at a particular time in the body [47,48]. Stable bio-based nanohydrogels have been produced and used for encapsulation and delivery of several bioactive compounds: such as ω-3 fatty acids [32], epigallocate-chin-3-gallate (the major catechin in green tea and a potent antioxidant) [33], α-tocopherol [34,49], and zinc [50] thus providing health-improving properties to beverages and food products.

The approaches more commonly used for the preparation of nanohydrogels are gelation processes [8,51]. Examples of nanosystems prepared by gelation are given in Table 2.1. Nanohydrogels permit overcoming some drawbacks inherent to other nanosystems, such as nanoemulsions or lipid nanoparticles in the preparation procedure and relatively low loading capacity. They can be produced with relatively low cost materials and do not require the addition of extra components. In general, their preparation method is simple, some examples of which are given in Table 2.1.

2.2.3 Nanoemulsions

Nanoemulsions consist of a lipid phase dispersed in an aqueous continuous phase (i.e., water or aqueous solution), where each oil droplet becomes surrounded by a thin interfacial layer of emulsifier molecules [52]. Usually, these systems have good stability against droplet aggregation because the range of attractive van der Waals forces acting between the droplets decreases with decreasing particle size. The particle size of nanoemulsions in food systems typically ranges from 10 to 100 nm [7]. In contrast, the range of steric repulsion is less dependent on particle size [53,54]. Nanoemulsions can act as carriers or delivery systems for lipophilic compounds, such as essential oils (i.e., ω-3-rich oils), polyphenolics (i.e., curcumin), antioxidants (i.e., quercetin), antimicrobials (i.e., thymol), and vitamins (i.e., vitamin A) [20,55]. Nanoemulsions not only can improve the solubility of lipophilic bioactive food components but also can improve their thermal stability and physiological performance by enhancing their bioavailability during the absorption in the GI tract. Furthermore, the impact of this encapsulation on the organoleptic properties of food products can be minimal [8,56,57]. It is important to point out that lipophilic, hydrophilic, and amphiphilic bioactive compounds can be incorporated within the oil droplets, the continuous phase, or the interfacial region of the oil-in-water in nanoemulsions, depending on their nature [58].

Nanoemulsions can be produced through high-energy methods, which need devices that use high mechanical energy input, such as high-pressure valve homogenizers, microfluidizers, and sonication methods, capable of generating intense disruptive forces that breakup the oil and water phases and lead to the formation of oil droplets. Low-energy approaches that require low energy for the production of nanoemulsions, and mainly depend on the intrinsic physico-chemical properties of surfactants and oily phase (i.e., phase inversion and solvent demixing methods) are also used [52].

2.2.4 Lipid nanoparticles

Solid lipid nanoparticles (SLNs) are oil-in-water emulsions in which the lipid phase has been either fully or partially solidified by a solid lipid, or a blend of solid lipids [59].

Examples of common lipids used in this type of nanosystems are sunflower and palm oils [60]. SLNs are usually created by preparing an oil-in-water nanoemulsion at a temperature above the melting point of the lipid phase, and subsequently cooled to temperatures below the crystallization point of the lipids [61,62].

Nanostructured lipid carriers (NLCs) are a new generation of SLNs consisting of a solid matrix where the overall solid content can be up to 95% of the total weight [63]. Some potential limitations typically associated with SLN are the relatively high water content of the dispersions (70%–99.9%) as it may lead to nanosystem instability; limited loading capacity; and issues in controlled release during storage. The loading capacity of SLNs is limited by the solubility of bioactive components in the lipid melt, the structure of the lipid matrix, and the polymeric state of the lipid matrix. In those cases where the lipid matrix consists of especially similar molecules (i.e., tristearin or tripalmitin), a perfect crystal with few imperfections is formed. As incorporated bioactive components are located between fatty acid chains, between the lipid layers, and also in crystal imperfections, a highly ordered crystal lattice cannot accommodate large amounts of components [64]. These weaknesses may be overcome by NLC [59]. Both systems, NLC and SLNs, represent suitable alternative carriers to liposomes and emulsions.

Lipid nanoparticles can be prepared with various techniques, such as high-pressure homogenization, microemulsion, sonication, nanoprecipitation, and emulsification-solvent evaporation [14,65]. Since nanoprecipitation and emulsification-solvent evaporation techniques have been explained earlier, only high-pressure homogenization, sonication, and microemulsion techniques will be described.

The production of these nanoparticles by high-pressure homogenization can be performed using either hot or cold homogenization techniques, in which the bioactive compound is dissolved, solubilized, or dispersed in the melted lipid in both approaches. In hot homogenization, the bioactive substance and lipid melt are dispersed in hot surfactant solution, both at the same temperature, and homogenized by high-speed stirring. The resulting preemulsion is then passed through a high-pressure homogenizer. In cold homogenization, the bioactive substance and lipid melt are cooled together and, after solidification, they are ground. This forms solid lipid microparticles that are homogenized in a cold surfactant solution and subjected to a high-pressure homogenizer, which results in the formation of SLN [66]. In the microemulsion technique, the lipid and aqueous phase are heated and homogenized at a temperature above the melting point of the lipid. The microemulsion follows dilution with cooled water, which leads to the breaking of the microemulsion into a nanoemulsion. The obtained nanoemulsion is cooled leading to the formation of SLN [67].

The production of nanoemulsions by sonication or by high-shear homogenization is carried out less commonly than by hot or cold homogenization. The lipid phase and aqueous phase are heated up to the same temperature and emulsified by mechanical stirring (high-shear homogenization) or sonication. The main disadvantage of these techniques is the presence of both micro- and nanoparticles in the final dispersion [68].

The high-pressure homogenization (hot process) has also many advantages in comparison to the other techniques as it allows an easy scale up and a short production time. It also avoids the use of organic solvents [59]. As an example, curcumin was successfully encapsulated in lipid nanoparticles [39] and its retention time and stability increased when assessed throughout in vivo studies [40].

2.2.5 Micelles

Micelles are aggregates of surfactants, such as caseins or whey proteins, of colloidal dimensions which are in equilibrium with their constituents. Additional components, named cosolvents or cosurfactants, may also be present in the aggregate and contribute to modify their dimensions and the interaction with the encapsulated substance. These systems are usually spherical (diameter <20 nm), with a hydrophobic core, composed of oil molecules and nonpolar surfactant chains. They have a hydrophilic shell composed of polar groups from surfactants. Lipophilic functional compounds can be encapsulated within the hydrophobic core of micelles, whereas hydrophilic bioactive food ingredients can be encapsulated in the core of reverse or inverted micelles, which possess polar inner groups and hydrophobic surface [69]. Micelles, which only scatter light weakly due to their small particle size (i.e., ranging from 2 to 20 nm), are suitable for encapsulating lipophilic compounds into clear foods and beverages [69]. Beta-casein, an amphiphilic self-assembling protein that can form micellar nanosystems, has been used to encapsulate and deliver hydrophobic therapeutic compounds, such as curcumin [42] and β-carotene [41].

The main challenges regarding the use of bio-based nanosystems for the encapsulation and controlled release of functional compounds include the following: (1) appropriate selection of the components to build the nanosystem due to the limited physicochemical properties inherent to food-grade materials; (2) their physical instability under environmental stresses (i.e., pH, salt, and heating); (3) limited control over oxidation of the functional compounds; and (4) ability to provide protection and stability for encapsulated compounds during hash food processing conditions or through GI passage [14]. The advances and recent strategies toward these limitations comprise the functionalization of the nanosystems through the use of combined polymers for their preparation or the creation of one or more layers of a polyelectrolyte surrounding the nanosystem. This is carried out through a layer-by-layer deposition technique, in which oppositely charged layers of polyelectrolytes are adsorbed on core materials with controllable thickness and properties [70]. This technique permits to add further layers by simple deposition of oppositely charged polyelectrolytes in solution, which promotes the adsorption of the polyelectrolytes on the top of the first layer. Repetition of this adsorption step leads to the formation of multilayers [71]. This strategy allows controlling the order and location of multiple polymer layers with nanoscale precision; and defining the concentrations of components encapsulated by simply varying the number of polyelectrolyte layers and selecting the polymer to be used from a wide range available, which increases the potential application of these nanosystems [72].

There is scarce information regarding the ability of micelles to maintain their properties when part of complex food matrices and resist real food processing conditions. Therefore it is of ultimate importance to understand and evaluate the influence of food matrices on the stability and bioavailability of the bioactive compounds in delivery systems, as well as the effect of the delivery systems on the physico-chemical and sensory properties of the final food product.

2.3 NANOTECHNOLOGY FOR FOOD SAFETY/SECURITY

Recent applications of nanotechnology in the food area include development of nanosized structures with the ability to improve the safety of food products [73]. An example of this contribution are sensors at the nanoscale that can support farmers by providing data on environmental conditions, fertilization, and pest control. This information can facilitate an early intervention and applying or decreasing the use of agrochemicals (i.e., antibiotics, pesticides, and nutrients) when necessary [1]. Moreover, nanoscale carriers can be used for the efficient delivery of fertilizers or pesticides and improving their effect. This is extremely relevant, given that the controlled use of these compounds can enhance the quality and safety of food product. For example, Siddiqui and Al-Whaibi [74] evaluated the benefits of applying silicon dioxide nanoparticles to tomato germinating seeds and significant effect on the characteristics of seed germination (percent of seed germination, mean germination time, seed germination index, seed vigor index, seedling fresh weight, and dry weight) was found [74]. Clay nanotubes have also been used as carriers of pesticides: their low cost, extended release, and better contact with plants allow reducing the amount of pesticide (around 80%), cost, and an impact of pesticide application on water streams [75].

Preserving food quality during storage and reducing food waste are main factors considered in food processes that transform agricultural products into foodstuffs for consumers [76]. During food processing, the control of the quality of the products in real time and the prompt identification of possible contaminations (pesticides, heavy metals, antibiotics, and pathogens) are the main challenges where nanotechnology can be very helpful [1]. The development of sensors with ability to quickly detect environmental changes during food storage (i.e., temperature, relative humidity, and oxygen exposure); physico-chemical modifications in food, such as color, pH; the presence of chemical species, such as nitrite or sulfite; or microbial contaminations, can be extremely useful to improve the safety and quality of food products.

Fresh food products that are either spoiled or exhibit unpalatable odors, color, or sensory properties are easily discerned by consumers. However food packaging obstructs the sensory perception of consumers and therefore consumers must rely on sell-by dates, which are determined by producers based on a set of idealized assumptions, among which is the way that the food will be stored or transported. Nanosensors provide a solution to this problem and inform on the quality of the food product. Developed nanosensors have been designed to be inserted in the packaging material, where they would serve as "noses" that can detect chemicals released during food spoilage [77]. Microfluidic devices are another efficient and high sensitivity type of nanosensors, with the ability to quickly detect pathogens in real time. Advantages of these fluidic sensors are their small dimension and their ability to detect relevant compounds rapidly in a small volume (few microliters), which has already led to widespread applications (such as medical, biological, and chemical analysis) [78].

Nanocantilevers are nanosensors composed of silicon-based materials. These present the ability to recognize proteins and detect pathogens and viruses. In fact, this capability enables the detection of biological binding interactions, such as antigen and antibody or enzyme and substrate, by physical and/or electromechanical signal. These nanosensors are also used in studies of molecular interactions and in the detection of toxins and antibiotic residues in food products [79]. Devices produced with the so-called *nanoelectromechanical systems technology* are also used for food quality control. They consist of advanced transducers which convert physico and chemical stimuli into electro signals which serve as tools in food preservation. They can control the storage environment and act as active "sell by" devices. An example of this device application is a digital transform spectrometer produced by Polychromix (Wilmington, MA, USA) which uses microelectromechanical systems technology to detect trans-fat content in foods [80]. Some food companies, such as Nano Engineered Applications [81], LamdaGen [82], and Nanolane [83] already make available nanotechnology products that help consumers to assess whether certain foods are safe. However, most applications of nanosensors for food-related components are still in their early development stages.

2.3.1 Detection of contaminants or adulterants, gases, and microrganisms

Many of the assays developed to detect molecular contaminants or adulterants in complex food matrices are based on observation of color changes or fluorescent emission that occurs in solutions containing nanoparticle and specific food contaminants [84]. Electrochemical detection based on immunoassays is also a useful methodology to detect organic molecules. It operates by binding antibodies to a conductive nanomaterial and detects conductivity changes when the antibodies bind the target contaminant. This type of detection is very selective and avoids possible interference caused by other food components.

Oxygen and moisture are relevant parameters to control food contamination given that the presence of high levels of oxygen and/or moisture leads to proliferation of microorganisms. Hence, continuous monitoring of the gas content in the food package headspace would also provide a means to assess the safety and quality of the food long after it has left the manufacturing facility. Noninvasive leak detection and methods for the determination of gas content are very valuable [85]. The use of nanostructures is being extremely useful to detect these parameters. Their high ratio of area/volume allows to detect faster and with more precision than the existing systems at bigger scale.

The presence of microorganisms in food products is one of the main causes of illnesses which, in extreme cases, can result in hospitalization. Therefore, it is important to determine the presence of bacteria, fungi, or viruses in food matrices that can cause food-borne diseases. Inexpensive methods for the fast and accurate detection of microbes in food are main goals of the food sector. Conventional methodologies for microbiological detection are based on immunological assays, which use selective antibody–antigen interactions. This approach is similar to the detection strategy used in nanomaterial-based microbial sensors. In contrast, nanomaterials can offer significant improvements in selectivity, speed, and sensitivity compared to conventional chemical or biological methods [86].

2.3.2 Removal of chemicals or pathogens from food

In contrast to the indirect detection systems to monitor food safety (i.e., nanosensors), there are methods that directly improve the safety of food products. Nanoparticles can be directly used to capture chemical contaminants or pathogenic microorganisms in food. For instance, Huang et al. [87] developed amine-functionalized magnetic nanoparticles to remove bacterial pathogens. These nanoparticles allowed rapid and efficient removal of bacteria from water and food matrices with high efficiency (88.5–99.1%) [87]. Strategies for water treatment involving nanotechnology are also being developed and improve food safety directly. These are discussed in Chapter 3, which is devoted to the application of nanotechnology in water treatment.

2.3.3 Food preparation surfaces

Surfaces containing nanomaterials, such as antimicrobial nanocoatings, can be used to create nonfouling surfaces for food preparation. These have the potential to reduce bacterial growth and decrease food production costs when reducing downtime caused by the clogging of processing machines. Different nanobased coatings are available for coating surfaces in contact with food in food preparation processes, and for coating machinery used in food processing, including nanosilica coating for hydrophobic self-cleaning surfaces; titanium dioxide or zinc oxide nanocoating for photocatalytic sterilization of food contact surfaces; and silver nanocoating for hygienic food preparation surfaces [88]. Table 2.2 shows some of examples of application of nanotechnology in these areas.

Table 2.2 Examples of application of nanotechnology in food safety

Area	Nanostructure	Purpose	References
Detection of organic molecules	Gold nanoparticles with cyanuric acid	Detection melanine	[89]
	Nanosensor	Detection gliadin	[90]
	Nanoscale liposome-based detector	Detection cyanide in water	[91]
	Gold nanoparticles and glucose sensitive enzymes	Measurement of glucose concentrations	[84]
Detection of gases	Photoactivated indicator based on nanosized TiO_2 or SnO_2 particles	Detection of oxygen	[92]
	Carbon-coated copper nanoparticles	Detection of moisture	[93]
	Fluorophore-encapsulated polymer nanobeads	Detection of carbon dioxide	[94]
Detection of microorganisms	Nanoscale magnetic particles attached to antibodies for *Listeria monocytogenes*	Separation of the target bacteria from artificially contaminated milk	[86]
	Nanoscale magnetic particles attached to antibodies for *Escherichia coli*	Separation of *Escherichia coli* from freshly ground beef	[86]

2.4 NANOTECHNOLOGY IN FOOD PROCESSING

Food processing is commonly used to modify the food matrix, maintain the nutritional quality of the food, and increase its shelf life. The food industry is seeking innovative and cheaper methods to produce and preserve food products, and nanotechnology can play an important role in the processing sector because it offers novel and effective tools for food manufacturing, encapsulation, and filtration [1]. Some of the most important applications of nanotechnology in the food-processing sector can be found in Table 2.3.

Table 2.3 Examples of application of nanotechnology in food processing

Purpose	Approaches	References
Nanofiltration	Selective passage of liquids on the basis of shape and size	[95]
Enzyme immobilization	Development of nanocarriers for enzymes with good stability, adaptability, and reusability	[96]
Texture improvement	Nanotubes as gelation and viscosifying agents	[97]
Color/flavor/aroma enhancers	Nanoencapsulation of color, flavor, or aroma enhancers	[98]
Meat replacers	Fibrillar protein aggregates as meat replacers	[99]

2.4.1 Nanofiltration

Nanofiltration consists of the separation of solutes above the molecular weight cut-off of the filter from the liquid medium. This method is applied in different sectors of food industry for product quality improvement and its application can be addressed either to isolated solutes with size above the cut-off of the filter, or to obtain a liquid free from certain solutes, such as in water treatment [100]. Although the separation mechanisms are still not completely known, they are determined by complex steric and electrical effects or electro neutrality principle, such as in the Donnan effect [101]. The membrane performance depends mainly on the effective pore radius and ratio of membrane thickness to porosity, as well as on the effective charge density [1]. Hence, membranes used in nanofiltration exhibit high permeability for salts (such as NaCl or KCl) and low permeability for bulkier compounds such as lactose and proteins [1].

2.4.2 Enzyme immobilization

Enzymes are commonly used in food processing methods to modify some food components and as a result, for instance, achieving enhanced flavor or higher nutritional value [102]. Nanocarriers can act as enzymes' support systems to provide improved activity; longer shelf life; and increase cost-effectiveness due to the better dispersion of enzymes (when attached to nanocarriers) in food matrices and large surface-to-volume ratios compared to traditional macroscale support materials [103]. Porcine pancreas lipase is a good example of successful enzyme immobilization carried out onto functionalized nanoscale SiO_2 spheres. It resulted in good stability, adaptability, and reusability of the immobilized enzyme [96]. Other authors showed that enzymatic activity and thermostability increased for lipase. Its denaturation and aggregation upon heating was effectively prevented when this enzyme was encapsulated in nanogels of cholesterol-bearing pullulan [104].

2.4.3 Texture improvement

Nanotubes are being investigated as gelation and viscosifying agents. For example, nanotubes based on milk protein α-lactalbumin present some advantages over conventional viscosifying and gelling agents: (1) They present high length-diameter ratio and provide stiffness, which make them efficient viscosifying agents. Less protein material is needed to increase viscosity of the food matrix as a result of their application. (2) Exhibit high storage moduli at low concentrations, which implies that less protein material is needed to form a gel. (3) Form transparent gels, which is attractive in food applications. (4) Their application makes possible reversible gel formation due to physical interactions. (5) Controlled disassembly of tubes is possible, which can be used to induce controlled reduction of gel strength [97].

2.4.4 Color/flavor/aroma enhancers

A technologically important food processing method with high economical potential involves the nanoencapsulation of food additives, such as flavors, colors, or aromas to enhance food attributes to suit consumers' preferences. It has been shown that nanoencapsulation can be used to protect volatile compounds in natural lipid mixtures (like roasted coffee oil). Furthermore, these nanocapsules could be used in turn as aroma enhancers for instant coffee [98].

2.4.5 Meat replacers

The replacement of meat products with products from alternative protein sources is an ambitious challenge. It implies creating meat-like structural and sensory properties with nonmeat protein-based ingredients following a complex optimization process (i.e., the formation of anisotropic structures at various length scales and in various concentration regimes [99]). Fibrillar protein aggregates are being developed as meat replacers, and nanotechnology has become one of the routes to construct fibrillar proteins with potential as food item [105].

The examples provided in this section give the picture of the diversity of applications that nanotechnology has in the food-processing sector. Although in its initial stage, it is expected that nanotechnology will have a high impact on food processing, for example, by making food safer or preserving/improving its nutritional value. In fact, all these applications of nanotechnology in the domain of food processing show promising results. However, due to its novelty, there are still many uncertainties that need to be clarified, such as the assessment of nanotoxicity and potential risks/benefits, before it can be widely applied in food industry.

2.5 PRODUCTS, MARKET, AND SOCIAL ACCEPTANCE

Agri-food companies are exploring the vast opportunities that nanotechnology can offer to sectors despite the limitations of current regulation (discussed in Chapter 9); from improvements in the nutritional value of foods, as explained in this work to improvements in food packaging (discussed in Chapters 7 and 8), food-processing, and primary production sectors. Nanotechnology is expected to reach an overall market of almost one trillion US dollars, while food encapsulation market is believed to reach 39 billion US dollars [10,106].

Nowadays, the number of inventions presented by companies in the nanotechnology field applied to foods is very important. The number of patents published in the last 14 years shows the economic importance in this field (Fig. 2.1B and Chapter 10). The interest of the industry for application of nanotechnology in food arose in 1970 with a patent on the production of an ultrathin polymer film with nanoporosity [107]. This invention was one of the cores of further inventions regarding inverse osmosis and nanofiltration. Subsequently, several companies presented inventions related to the use of nanostructured membranes for the purification and filtration of water [108–110]: these membranes are widely used nowadays in food processing and water purification.

The current state of intellectual property in the food sector has been thoroughly discussed in Chapter 10; we will only highlight here examples relevant to the area discussed in this chapter with historical view. In 1979, Professor Rosano published a patent describing the production of a microemulsion with possible applications in food products [111]. However, it was not until 1986 that BASF (USA) showed interest in microemulsions in foods with the publication of a patent entitled "Microemulsion" claiming their use in foods, among other applications [112]. In 1990, a patent submitted by the Rohm And Haas Company (company bought in 2009 by Dow Chemical Company) protected a microbiocidal microemulsion (sizes below 100 nm) and its applications in food processing, and its inclusion in dairy and animal products [113]. In the same year, Procter & Gamble Company was granted a patent entitled "Food microemulsion" which covered several applications [114]. Also, in the early 1990s the company Coletica (France) developed a patent that protected a method for producing nanocapsules for food applications [115]; and the Southern Research Institute (AL, USA) presented an invention on the production of capsules at micro and nanoscale for encapsulation of food ingredients [116]. Since then, a great number of inventions concerning encapsulation of compounds through the use of nanocapsules, nanoemulsions, or nanotubes have been presented [117,118].

Pioneer work on the inclusion of nanotechnology in packaging materials was carried out by the company AlliedSignal (that joined Honeywell in 1999). They presented polymer nanocomposites synthetized by a melting process, and claimed that an exfoliated layered material derivatized with reactive organo silanes had applications for the fabrication of extruded and laminated film for use in food packaging [119]. The publication of a patent regarding the use of a metal ion, such as silver, zinc, or copper, with size between 0.1 and 10 μm, in a resin, showed the possibility of using nanoparticles to produce packaging materials with antibacterial properties in the late 1990s [120].

It is expected that a great number of products containing nanotechnology will arrive at the market in the next few years. Table 2.4 presents examples of commercial applications already available in the market. However, it is important to mention that it is expected that several other products in the market may contain nanotechnology despite the corresponding owner companies not claiming the inclusion of nanotechnology in them.

The will to protect nanotechnology applications, such as the inventions cited that were protected 30 years ago, and also the most recent developments patented (some

Table 2.4 Commercial applications of nanotechnology products established in the agri-food industry

Product	Agri-food sector	Nanostructure type	Company	Country	References
Canoila active oil	Nutrition	Nanoemulsion	Shemen Industries	Israel	[121,122]
NovaSol® beverages	Nutrition	Nanoemulsion	Aquanova	Germany	[123]
Nutralease beverages	Nutrition	Nanosized self-assembled structured liquids	Nutralease Ltd.	Israel	[124]
SunActive®	Nutrition	Nanoemulsions	Tayo International	Japan	[125]
Colors from Nature®	Nutrition	Nanoemulsions	Wild Flavours Inc	USA	[126]
NanoResveratrol®	Nutrition	Nanoemulsions	Life Enhancement	USA	[127]
SprayforLife	Nutrition	Nanoemulsions	NanoSynergy Worldwide	USA	[128]
Nutri-Nano	Nutrition	Nanoemulsions	Solgar	USA	[129]
Altrient	Nutrition	Liposomes	Livon Laboratories	USA	[130]
Bioral	Nutrition		Bioral Nutrient Delivery, LLC	USA	[131]
BioNutriCoat	Nutrition/Food packaging	Nanoemulsions	Improveat	Portugal	[132]
Fresh Box	Food packaging	Silver nanoparticles	BlueMoonGoods, LLC	USA	[133]
N-Coat	Food packaging	Clay nanoparticles	Multifilm Packaging Corporation	USA	[134]
FresherLonger Miracle Food Storage	Food packaging	Silver nanoparticles	Sharper Image	USA	[135]
Nano Silver Baby Mug Cup	Food packaging	Silver nanoparticles	Baby Dream® Co., Ltd.	Korea	[136]
Imperm	Food packaging	Clay nanoparticles	Nanocor®	USA	[137]
BactiBlock	Food packaging	Silver-ion based technology in a nanoclay carrier	NanoBioMatters	Spain	
Nansulate	Food processing	Nanocomposite	Industrial Nanotech Inc.	USA	[138]
Bioni nano paint	Food processing	Silver nanoparticles	Bioni	Germany	[139]
Nano air-filter	Food processing	Silver nanoparticles	SongSing Nano Technology Co., Ltd.	Taiwan	[140]
Wifi nanosensors	Primary production	Nanosensors	Accenture	USA	[141]
Primo Maxx	Primary production	Nanoemulsions	Syngenta	Switzerland	[141]

examples compiled in Table 2.4) shows awareness that nanotechnology could and can trigger new market opportunities. Indeed, the advantages offered by nanotechnology in innovative food products can lead to market differentiation and economical gains [10,141,142]. Today, nanotechnology is considered as a major tool to help to address key societal challenges by the European Union. For this reason, it is included in the EU Framework Programme for Research and Innovation, the Horizon 2020. One of the objectives of research calls from this program is to reduce the gap between research in nanotechnology and finalized products reaching market, while contributing to economical sustainable growth, competitiveness of smaller companies, promoting specialized jobs and, at the same time, improve quality of life [143]. Therefore, to address these challenges, the EU provides funds to scale-up laboratory prototypes to industrial scale and establish the viability of these new manufacturing technologies. The developed food products are aimed to be commercialized as multifunctional, economic and environmentally friendly, and safety will need to be assessed in terms of potential impact on human and environment health [143]. Some specific technologies developed under the EU research framework are shown in Chapter 3. The EU seeks to support competitiveness by accelerating market acceptance of nanolabeled products. This will be achieved by continuous improvement in manufacturing processes and by defining regulatory strategies [143].

The potential of nanotechnology for solving societal challenges, such as food security, nutrition, food safety, and environmental protection is also recognized by the US Department of Agriculture, which is funding nine research projects (with budget of US $3.8 million) to support solutions to the societal challenges based on nanotechnology [144].

Despite all the potential benefits that nanotechnology could bring to the agri-food industry, some risks need to be assessed regarding the safety workers manipulating nanomaterials, potential side effects to human health, such as oxidative damage, inflammation of the GI system and cancers, as well as lesions of liver and kidneys due to acute toxic responses, need to be investigated [1,10]. Additionally, consequences from the introduction of nanoparticles in the environment need to be measured.

Recently, some work regarding the awareness and acceptability of consumers to purchase food products developed using nanotechnology has been published: these give an insight on how the consumers react. For instance, Roosen and coworkers analyzed the trust and willingness of consumers to acquire food products developed with nanotechnology by means of online surveys in Canada and Germany. These authors also analyzed the acceptance of nanobased functional food and changes in the trust of the consumers when they learnt that certain properties in the product were achieved with nanotechnology (such as enrichment with vitamins or other chemicals that increase food protection) [145]. These studies showed that the use of nanotechnology in food products is a real concern for consumers. However, it was unclear if the consumers' concerns were related to the lack of information regarding nanotechnology in general, or lack of knowledge on the uses of nanotechnology in the food industry specifically [145]. Additionally, these authors conclude that the use of nanotechnology is considered "bad news," with this being consistent among Canadian and German consumers [145]. The consumers' concern was also expressed in studies performed in the United States and Switzerland [145].

Siegrist and coworkers identified factors that can influence the acceptability of nanotechnology in food products through questionnaires. In general, participants were hesitant to buy nanotechnology-based products; either foods or food with packaging using nanotechnology. With all, nanotechnology in packaging is seen as more beneficial than nanotechnology applied directly on foods. The questionnaires' results suggested that the social trust has a major impact on the benefits and observed risks, whereas the perceived benefits appear to be an important predictor factor for the acceptability and willingness of consumers to buy food products manufactured with the use of nanotechnology [146]. Zhou [147] achieved similar conclusions as in the previous study: observing that consumers are more willing to accept nanotechnology when they are aware of the benefits added to the food product [147]. In another study, Handford and coworkers surveyed the awareness and attitudes to agri-food organizations in Ireland regarding the application of nanotechnology through interviews (carried out in 14 organizations) and online questionnaire (carried out in 88 organizations). The study identified that the information on nanotechnology applications provided by the food industry is low and responses were neither positive nor negative. Also, the key benefits pointed out regarding the use of food products containing nanotechnology were safer food products, reduced waste, and increased product shelf life. As key barriers to the full implementation of nanotechnology, the study identified the unknown hazards to human health and environmental impacts, consumer approval, and media framing [148]. Given the overall agreement among the results from independent studies carried out in different countries, scientists, governmental entities, and agri-food industry need to cooperate to increase the understanding, awareness, and acceptability of nanotechnology by media, and more importantly, by consumers. All the involved parties should clarify the safety of nanotechnology food products and applications, addressing the potential benefits associated with its consumption, while implementing legislative framework for the application of nanotechnology in the agri-food industry.

ACKNOWLEDGMENTS

Ana C. Pinheiro, Hélder D. Silva, Ana I. Bourbon, Óscar L. Ramos (SFRH/BPD/101181/2014, SFRH/BD/81288/2011, SFRH/BD/73178/2010, and SFRH/BPD/80766/2011, respectively) are recipients of a fellowship from the Fundação para a Ciência e Tecnologia (FCT, POPH-QREN, and FSE Portugal).

REFERENCES

[1] N. Dasgupta, S. Ranjan, D. Mundekkad, C. Ramalingam, R. Shanker, A. Kumar, Nanotechnology in agro-food: from field to plate, Food Res. Int. 69 (2015) 381–400.

[2] M. Cerqueira, A. Pinheiro, H. Silva, P. Ramos, M. Azevedo, M. Flores-López, M. Rivera, A. Bourbon, Ó. Ramos, A. Vicente, Design of bio-nanosystems for oral delivery of functional compounds, Food Eng. Rev. 6 (1-2) (2014) 1–19.

[3] L. Chen, G.E. Remondetto, M. Subirade, Food protein-based materials as nutraceutical delivery systems, Trends Food Sci. Technol. 17 (5) (2006) 272–283.

[4] P. de Vos, M.M. Faas, M. Spasojevic, J. Sikkema, Encapsulation for preservation of functionality and targeted delivery of bioactive food components, Int. Dairy J. 20 (4) (2010) 292–302.

[5] C.-F. Chau, S.-H. Wu, G.-C. Yen, The development of regulations for food nanotechnology, Trends Food Sci. Technol. 18 (5) (2007) 269–280.

[6] V. Nedovic, A. Kalusevic, V. Manojlovic, S. Levic, B. Bugarski, An overview of encapsulation technologies for food applications, Procedia Food Sci. 1 (2011) 1806–1815.

[7] D.J. McClements, E.A. Decker, Y. Park, J. Weiss, Structural design principles for delivery of bioactive components in nutraceuticals and functional foods, Crit. Rev. Food Sci. Nutr. 49 (6) (2009) 577–606.

[8] M.A. Cerqueira, A.C. Pinheiro, H.D. Silva, P.E. Ramos, M.A. Azevedo, M.L. Flores-López, M.C. Rivera, A.I. Bourbon, O.L. Ramos, A.A. Vicente, Design of bio-nanosystems for oral delivery of functional compounds, Food Eng. Rev. 6 (2014) 1–19.

[9] F.M. Hilty, M.B. Zimmermann, Nano-structured minerals and trace elements for food and nutrition applications, in: H.-S. Kwak (Ed.), Nano- and Microencapsulation for Foods, Wiley-Blackwell, Chichester, West Sussex, UK, 2014, pp. 199–222.

[10] S. Neethirajan, D. Jayas, Nanotechnology for the food and bioprocessing industries, Food Bioprocess. Technol. 4 (1) (2011) 39–47.

[11] C. Wandrey, A. Bartkowiak, S.E. Harding, Materials for encapsulation, in: N.J. Zuidam, V.A. Nedovic (Eds.), Encapsulation Technologies for Food Active Ingredients and Food Processing, Springer, Dordrecht, The Netherlands, 2009, pp. 31–100.

[12] H.-S. Kwak, Overview of Nano- and Microencapsulation for Foods, in: H.-S. Kwak (Ed.), Nano- and Microencapsulation for Foods, Wiley-Blackwell, Chichester, West Sussex, UK, 2014, pp. 1–14.

[13] D.J. McClements, Nanoparticle- and Microparticle-Based Delivery Systems: Encapsulation, Protection and Release of Active Components, CRC Press, Boca Raton, FL, (2014).

[14] I.J. Joye, G. Davidov-Pardo, D.J. McClements, Nanotechnology for increased micronutrient bioavailability, Trends Food Sci. Technol. 40 (2) (2014) 168–182.

[15] D.J. McClements, J. Rao, Food-grade nanoemulsions: formulation, fabrication, properties, performance, biological fate, and potential toxicity, Crit. Rev. Food Sci. Nutr. 51 (4) (2011) 285–330.

[16] M.A. Augustin, Y. Hemar, Nano- and micro-structured assemblies for encapsulation of food ingredients, Chem. Soc. Rev. 38 (4) (2009) 902–912.

[17] O.G. Jones, D.J. McClements, Functional biopolymer particles: design, fabrication, and applications, Comprehens. Rev. Food Sci. Food Saf. 9 (4) (2010) 374–397.

[18] P.N. Ezhilarasi, P. Karthik, N. Chhanwal, C. Anandharamakrishnan, Nanoencapsulation techniques for food bioactive components: a review, Food Bioprocess. Technol. 6 (3) (2013) 628–647.

[19] M.A. Augustin, P. Sanguansri, Nanostructured materials in the food industry, in: L.T. Steve (Ed.), Advances in Food, Nutrition Research, Academic Press, London, UK, 2009, pp. 183–213.

[20] P. Sanguansri, M.A. Augustin, Nanoscale materials development—a food industry perspective, Trends Food Sci. Technol. 17 (10) (2006) 547–556.

[21] M. Aghbashlo, H. Mobli, A. Madadlou, S. Rafiee, Integrated optimization of fish oil microencapsulation process by spray drying, J. Microencapsul. 29 (8) (2012) 790–804.

[22] S.M. Jafari, E. Assadpoor, B. Bhandari, Y. He, Nano-particle encapsulation of fish oil by spray drying, Food Res. Int. 41 (2) (2008) 172–183.

[23] M. Fathi, Á. Martín, D.J. McClements, Nanoencapsulation of food ingredients using carbohydrate based delivery systems, Trends Food Sci. Technol. 39 (1) (2014) 18–39.

[24] C.P. Reis, R.J. Neufeld, A.J. Ribeiro, F. Veiga, Nanoencapsulation I. Methods for preparation of drug-loaded polymeric nanoparticles, nanomedicine: nanotechnology, Biol. Med. 2 (1) (2006) 8–21.

[25] T.S.H. Leong, T.J. Wooster, S.E. Kentish, M. Ashokkumar, Minimising oil droplet size using ultrasonic emulsification, Ultrason. Sonochem. 16 (6) (2009) 721–727.
[26] M.J. Choi, S. Briançon, J. Andrieu, S.G. Min, H. Fessi, Effect of freeze-drying process conditions on the stability of nanoparticles, Drying Technol. 22 (1-2) (2004) 335–346.
[27] S.S. Guterres, R.C.R. Beck, A.R. Pohlmann, Spray-drying technique to prepare innovative nanoparticulated formulations for drug administration: a brief overview, Braz. J. Phys. 39 (2009) 205–209.
[28] F.M. Goycoolea, G. Lollo, C. Remuñán-López, F. Quaglia, M.J. Alonso, Chitosan-alginate blended nanoparticles as carriers for the transmucosal delivery of macromolecules, Biomacromolecules 10 (7) (2009) 1736–1743.
[29] M.d.S. Silva, D.S. Cocenza, R. Grillo, N.F.S.d. Melo, P.S. Tonello, L.C.d. Oliveira, D.L. Cassimiro, A.H. Rosa, L.F. Fraceto, Paraquat-loaded alginate/chitosan nanoparticles: preparation, characterization and soil sorption studies, J. Hazard. Mater. 190 (1-3) (2011) 366–374.
[30] R.K. Das, N. Kasoju, U. Bora, Encapsulation of curcumin in alginate-chitosan-pluronic composite nanoparticles for delivery to cancer cells, nanomedicine: nanotechnology, Biol. Med. 6 (1) (2010) 153–160.
[31] M.A. Azevedo, A.I. Bourbon, A.A. Vicente, M.A. Cerqueira, Alginate/chitosan nanoparticles for encapsulation and controlled release of vitamin B2, Int. J. Biol. Macromol. 71 (2014) 141–146.
[32] P. Zimet, Y.D. Livney, Beta-lactoglobulin and its nanocomplexes with pectin as vehicles for ω-3 polyunsaturated fatty acids, Food Hydrocolloids 23 (4) (2009) 1120–1126.
[33] B. Li, W. Du, J. Jin, Q. Du, Preservation of (−)-epigallocatechin-3-gallate antioxidant properties loaded in heat treated β-lactoglobulin nanoparticles, J. Agric. Food Chem. 60 (13) (2012) 3477–3484.
[34] W. Somchue, W. Sermsri, J. Shiowatana, A. Siripinyanond, Encapsulation of α-tocopherol in protein-based delivery particles, Food Res. Int. 42 (8) (2009) 909–914.
[35] A.H. Sneharani, J.V. Karakkat, S.A. Singh, A.G.A. Rao, Interaction of curcumin with β-lactoglobulin—stability, spectroscopic analysis, and molecular modeling of the complex, J. Agric. Food Chem. 58 (20) (2010) 11130–11139.
[36] A.C. Pinheiro, M. Lad, H.D. Silva, M.A. Coimbra, M. Boland, A.A. Vicente, Unravelling the behaviour of curcumin nanoemulsions during in vitro digestion: effect of the surface charge, Soft Matter 9 (11) (2013) 3147–3154.
[37] H.D. Silva, M.A. Cerqueira, B.W.S. Souza, C. Ribeiro, M.C. Avides, M.A.C. Quintas, J.S.R. Coimbra, M.G. Carneiro-da-Cunha, A.A. Vicente, Nanoemulsions of β-carotene using a high-energy emulsification–evaporation technique, J. Food Eng. 102 (2) (2011) 130–135.
[38] A. Noack, G. Hause, K. Mäder, Physicochemical characterization of curcuminoid-loaded solid lipid nanoparticles, Int. J. Pharm. 423 (2) (2012) 440–451.
[39] V. Kakkar, S. Singh, D. Singla, I.P. Kaur, Exploring solid lipid nanoparticles to enhance the oral bioavailability of curcumin, Mol. Nutr. Food Res. 55 (3) (2011) 495–503.
[40] A. Guri, I. Gulseren, M. Corredig, Utilization of solid lipid nanoparticles for enhanced delivery of curcumin in cocultures of HT29-MTX and Caco-2 cells, Food Funct. 4 (9) (2013) 1410–1419.
[41] M.-J. Sáiz-Abajo, C. González-Ferrero, A. Moreno-Ruiz, A. Romo-Hualde, C.J. González-Navarro, Thermal protection of β-carotene in re-assembled casein micelles during different processing technologies applied in food industry, Food Chem. 138 (2-3) (2013) 1581–1587.
[42] M. Esmaili, S.M. Ghaffari, Z. Moosavi-Movahedi, M.S. Atri, A. Sharifizadeh, M. Farhadi, R. Yousefi, J.-M. Chobert, T. Haertlé, A.A. Moosavi-Movahedi, Beta casein-micelle as a nano vehicle for solubility enhancement of curcumin; food industry application, LWT-Food Sci. Technol. 44 (10) (2011) 2166–2172.
[43] J. Martins, Ó. Ramos, A. Pinheiro, A. Bourbon, H. Silva, M. Rivera, M. Cerqueira, L. Pastrana, F.X. Malcata, Á. González-Fernández, A. Vicente, Edible bio-based nanostructures: delivery, absorption and potential toxicity, Food Eng. Rev. 7 (4) (2015) 491–513.
[44] M. Rajaonarivony, C. Vauthier, G. Couarraze, F. Puisieux, P. Couvreur, Development of a new drug carrier made from alginate, J. Pharm. Sci. 82 (9) (1993) 912–917.
[45] Z. Fang, B. Bhandari, Encapsulation of polyphenols—a review, Trends Food Sci. Technol. 21 (10) (2010) 510–523.
[46] J.P. Rao, K.E. Geckeler, Polymer nanoparticles: preparation techniques and size-control parameters, Prog. Polym. Sci. 36 (7) (2011) 887–913.

[47] O.L. Ramos, R.N. Pereira, A. Martins, R. Rodrigues, C. Fuciños, J.A. Teixeira, L. Pastrana, F.X. Malcata, A.A. Vicente, Design of whey protein nanostructures for incorporation and release of nutraceutical compounds in food, Crit. Rev. Food Sci. Nutr. (2015) doi:10.1080/10408398.2014.993749.
[48] F. Liu, M.W. Urban, Recent advances and challenges in designing stimuli-responsive polymers, Prog. Polym. Sci. 35 (1-2) (2010) 3–23.
[49] P. Relkin, R. Shukat, Food protein aggregates as vitamin-matrix carriers: impact of processing conditions, Food Chem. 134 (4) (2012) 2141–2148.
[50] İ. Gülseren, Y. Fang, M. Corredig, Zinc incorporation capacity of whey protein nanoparticles prepared with desolvation with ethanol, Food Chem. 135 (2) (2012) 770–774.
[51] A.I. Bourbon, A.C. Pinheiro, M.G. Carneiro-da-Cunha, R.N. Pereira, M.A. Cerqueira, A.A. Vicente, Development and characterization of lactoferrin-GMP nanohydrogels: evaluation of pH, ionic strength and temperature effect, Food Hydrocolloids 48 (0) (2015) 292–300.
[52] H. Silva, M. Cerqueira, A. Vicente, Nanoemulsions for food applications: development and characterization, Food Bioprocess Technol. 5 (3) (2012) 854–867.
[53] D. McClements, Food Emulsions: Principles, Practice, and Techniques, CRC Press, Boca Raton, FL, (2005).
[54] T.F. Tadros, Emulsion Formation, Stability, and Rheology, Wiley-VCH Verlag GmbH & Co. KGaA, Weinheim, Germany, (2013) p. 1–75.
[55] A.C. Pinheiro, M.A. Coimbra, A.A. Vicente, In vitro behaviour of curcumin nanoemulsions stabilized by biopolymer emulsifiers—Effect of interfacial composition, Food Hydrocolloids 52 (2016) 460–467.
[56] G. Spigno, F. Donsì, D. Amendola, M. Sessa, G. Ferrari, D.M. De Faveri, Nanoencapsulation systems to improve solubility and antioxidant efficiency of a grape marc extract into hazelnut paste, J. Food Eng. 114 (2) (2013) 207–214.
[57] E. Acosta, Bioavailability of nanoparticles in nutrient and nutraceutical delivery, Curr. Opin. Colloid Interface Sci. 14 (1) (2009) 3–15.
[58] C.C. Berton-Carabin, J.N. Coupland, R.J. Elias, Effect of the lipophilicity of model ingredients on their location and reactivity in emulsions and solid lipid nanoparticles, Colloids Surf. A 431 (2013) 9–17.
[59] J. Pardeike, A. Hommoss, R.H. Müller, Lipid nanoparticles (SLN, NLC) in cosmetic and pharmaceutical dermal products, Int. J. Pharm. 366 (1-2) (2009) 170–184.
[60] M. Sessa, A. Casazza, P. Perego, R. Tsao, G. Ferrari, F. Donsì, Exploitation of polyphenolic extracts from grape marc as natural antioxidants by encapsulation in lipid-based nanodelivery systems, Food Bioprocess. Technol. 6 (10) (2013) 2609–2620.
[61] D.J. McClements, E.A. Decker, J. Weiss, Emulsion-based delivery systems for lipophilic bioactive components, J. Food Sci. 72 (8) (2007) R109–R124.
[62] J. Weiss, E. Decker, D.J. McClements, K. Kristbergsson, T. Helgason, T. Awad, Solid lipid nanoparticles as delivery systems for bioactive food components, Food Biophys. 3 (2) (2008) 146–154.
[63] J.-Y. Fang, C.-L. Fang, C.-H. Liu, Y.-H. Su, Lipid nanoparticles as vehicles for topical psoralen delivery: solid lipid nanoparticles (SLN) versus nanostructured lipid carriers (NLC), Eur. J. Pharm. Biopharm. 70 (2) (2008) 633–640.
[64] S. Mukherjee, S. Ray, R.S. Thakur, Solid lipid nanoparticles: a modern formulation approach in drug delivery system, Ind. J. Pharm. Sci. 71 (4) (2009) 349–358.
[65] J.S. Negi, P. Chattopadhyay, A.K. Sharma, V. Ram, Development of solid lipid nanoparticles (SLNs) of lopinavir using hot self nano-emulsification (SNE) technique, Eur. J. Pharm. Sci. 48 (1-2) (2013) 231–239.
[66] R.H. Müller, M. Radtke, S.A. Wissing, Solid lipid nanoparticles (SLN) and nanostructured lipid carriers (NLC) in cosmetic and dermatological preparations, Adv. Drug Deliv. Rev. 54 (Suppl.) (2002) S131–S155.
[67] W. Tiyaboonchai, W. Tungpradit, P. Plianbangchang, Formulation and characterization of curcuminoids loaded solid lipid nanoparticles, Int. J. Pharm. 337 (1-2) (2007) 299–306.
[68] V.R. Sinha, S. Srivastava, H. Goel, V. Jindal, Solid lipid nanoparticles (SLNs)—trends and implications in drug targeting, Int. J. Adv. Pharm. Sci. 1 (2011) 212–238.
[69] L. Sagalowicz, M.E. Leser, Delivery systems for liquid food products, Curr. Opin. Colloid Interface Sci. 15 (1-2) (2010) 61–72.

[70] A.P.R. Johnston, G.K. Such, S.J. Dodds, F. Caruso, Engineered Layer-by-Layer Assembled Capsules for Biomedical Applications, Wiley-VCH Verlag GmbH & Co. KGaA, Weinheim, Germany, (2012) p. 801–829.
[71] M.M. de Villiers, D.P. Otto, S.J. Strydom, Y.M. Lvov, Introduction to nanocoatings produced by layer-by-layer (LbL) self-assembly, Adv. Drug Deliv. Rev. 63 (9) (2011) 701–715.
[72] S. De Koker, L.J. De Cock, P. Rivera-Gil, W.J. Parak, R. Auzély Velty, C. Vervaet, J.P. Remon, J. Grooten, B.G. De Geest, Polymeric multilayer capsules delivering biotherapeutics, Adv. Drug Deliv. Rev. 63 (9) (2011) 748–761.
[73] Q. Chaudhry, R. Watkins, L. Castle, Nanotechnologies in the food arena: new opportunities, new questions, new concerns, in: Q. Chaudhry, L. Castle, R. Watkins (Eds.), Nanotechnologies in Food, RSC Nanoscience & Nanotechnology, Royal Society of Chemistry, Cambridge, UK, 2010, pp. 1–17.
[74] M.H. Siddiqui, M.H. Al-Whaibi, Role of nano-SiO_2 in germination of tomato (Lycopersicum esculentum seeds Mill.), Saudi J. Biol. Sci. 21 (1) (2014) 13–17.
[75] K. Murphy, Nanotechnology: Agriculture's Next "Industrial" Revolution, Spring, Editor. 2008.
[76] V. Vamvakaki, N.A. Chaniotakis, Pesticide detection with a liposome-based nano-biosensor, Biosens. Bioelectron. 22 (12) (2007) 2848–2853.
[77] N. Sozer, J.L. Kokini, Nanotechnology and its applications in the food sector, Trends Biotechnol. 27 (2) (2009) 82–89.
[78] T. Vo-Dinh, B.M. Cullum, D.L. Stokes, Nanosensors and biochips: frontiers in biomolecular diagnostics, Sensors Actuators B 74 (1-3) (2001) 2–11.
[79] J. Fritz, Cantilever biosensors, Analyst 133 (7) (2008) 855–863.
[80] S.K. Ritter, An eye on food, Chem. Eng. News 83 (27) (2005) 28–34.
[81] Available from: http://www.neapplications.com
[82] Available from: http://lamdagen.com
[83] Available from: http://www.nano-lane.com
[84] T.V. Duncan, Applications of nanotechnology in food packaging and food safety: barrier materials, antimicrobials and sensors, J. Colloid Interface Sci. 363 (1) (2011) 1–24.
[85] M. Smolander, E. Hurme, R. Ahvenainen, Leak indicators for modified-atmosphere packages, Trends Food Sci. Technol. 8 (4) (1997) 101–106.
[86] N. Sanvicens, C. Pastells, N. Pascual, M.P. Marco, Nanoparticle-based biosensors for detection of pathogenic bacteria, TrAC Trends Anal. Chem. 28 (11) (2009) 1243–1252.
[87] Y.-F. Huang, Y.-F. Wang, X.-P. Yan, Amine-functionalized magnetic nanoparticles for rapid capture and removal of bacterial pathogens, Environ. Sci. Technol. 44 (20) (2010) 7908–7913.
[88] Q. Chaudhry, L. Castle, Food applications of nanotechnologies: an overview of opportunities and challenges for developing countries, Trends Food Sci. Technol. 22 (11) (2011) 595–603.
[89] K. Ai, Y. Liu, L. Lu, Hydrogen-bonding recognition-induced color change of gold nanoparticles for visual detection of melamine in raw milk and infant formula, J. Am. Chem. Soc. 131 (27) (2009) 9496–9497.
[90] M. Staiano, E.G. Matveeva, M. Rossi, R. Crescenzo, Z. Gryczynski, I. Gryczynski, L. Iozzino, I. Akopova, S. D'Auria, Nanostructured silver-based surfaces: new emergent methodologies for an easy detection of analytes, ACS Appl. Mater. Interfaces 1 (12) (2009) 2909–2916.
[91] Y. Liu, K. Ai, X. Cheng, L. Huo, L. Lu, Gold-nanocluster-based fluorescent sensors for highly sensitive and selective detection of cyanide in water, Adv. Funct. Mater. 20 (6) (2010) 951–956.
[92] S.K. Lee, A. Mills, A. Lepre, An intelligence ink for oxygen, Chem. Commun. 10 (17) (2004) 1912–1913.
[93] N.A. Luechinger, S. Loher, E.K. Athanassiou, R.N. Grass, W.J. Stark, Highly sensitive optical detection of humidity on polymer/metal nanoparticle hybrid films, Langmuir 23 (6) (2007) 3473–3477.
[94] C. Von Bültzingslöwen, A.K. McEvoy, C. McDonagh, B.D. MacCraith, I. Klimant, C. Krause, O.S. Wolfbeis, Sol-gel based optical carbon dioxide sensor employing dual luminophore referencing for application in food packaging technology, Analyst 127 (11) (2002) 1478–1483.
[95] D. Zhou, L. Zhu, Y. Fu, M. Zhu, L. Xue, Development of lower cost seawater desalination processes using nanofiltration technologies—a review, Desalination 376 (2015) 109–116.
[96] Y.-X. Bai, Y.-F. Li, Y. Yang, L.-X. Yi, Covalent immobilization of triacylglycerol lipase onto functionalized nanoscale SiO_2 spheres, Process Biochem. 41 (4) (2006) 770–777.
[97] J.F. Graveland-Bikker, C.G. de Kruif, Unique milk protein based nanotubes: food and nanotechnology meet, Trends Food Sci. Technol. 17 (5) (2006) 196–203.

[98] E.B. Freiberger, K.C. Kaufmann, E. Bona, P.H. Hermes de Araújo, C. Sayer, F.V. Leimann, O.H. Gonçalves, Encapsulation of roasted coffee oil in biocompatible nanoparticles, LWT-Food Sci. Technol. 64 (1) (2015) 381–389.
[99] J.M. Manski, A.J. van der Goot, R.M. Boom, Formation of fibrous materials from dense calcium caseinate dispersions, Biomacromolecules 8 (4) (2007) 1271–1279.
[100] C.M. Lopes, J.R. Fernandes, P. Martins-Lopes, Application of nanotechnology in the agro-food sector, Food Technol. Biotechnol. 51 (2) (2013) 183–197.
[101] X. Xu, H.G. Spencer, Transport of electrolytes through a weak acid nanofiltration membrane: effects of flux and crossflow velocity interpreted using a fine-porous membrane model, Desalination 113 (1) (1997) 85–93.
[102] P. Fernandes, Enzymes in food processing: a condensed overview on strategies for better biocatalysts, Enzyme Res. 2010 (2010) 19.
[103] L. Rashidi, K. Khosravi-Darani, The applications of nanotechnology in food industry, Crit. Rev. Food Sci. Nutr. 51 (8) (2011) 723–730.
[104] S.-i. Sawada, K. Akiyoshi, Nano-encapsulation of lipase by self-assembled nanogels: induction of high enzyme activity and thermal stabilization, Macromol. Biosci. 10 (4) (2010) 353–358.
[105] D. Coles, L.J. Frewer, Nanotechnology applied to European food production—a review of ethical and regulatory issues, Trends Food Sci. Technol. 34 (1) (2013) 32–43.
[106] Global, Industry, and Analysts, Food Encapsulation: A Global Strategic Business Report, 2010. Available from: http://www.ift.org/food-technology/daily-news/2010/july/07/global-food-encapsulation-market-to-reach-$39b-by-2015.aspx
[107] R.H. Forester, P.S. Francis, Method of producing an ultrathin polymer film laminate, U.S. Patent, Editor. North Star Research & Development Institute, USA, 1970.
[108] D. Berg, D. Ciliberti, L. Scala, Interface condensation desalination membranes, U.S. Patent, Editor. Westinghouse Electric Corporation, USA, 1973.
[109] J.E. Cadotte, D.R. Walker, Novel water softening membranes, U.S. Patent, Editor. Filmtec Corporation, USA, 1989.
[110] R.E. Kesting, Microporous polyester membranes and polymer assisted phase inversion process for their manufacture, U.S. Patent, Editor. Chemical Systems Incorporated, USA, 1976.
[111] H.L. Rosano, Method for preparing microemulsions, U.S. Patent, Editor. Rosano, H.L., USA, 1979.
[112] B. Thir, E.M. Dexheimer, Microemulsions, U.S. Patent, Editor. Basf Wyandotte Corporation, USA, 1986.
[113] J.R. Mattox, Microbicidal microemulsion, U.S. Patent, Editor. Rohm And Haas Company, USA, 1990.
[114] M. El-Nokaly, G.D. Hiler, J. McGrady, Food microemulsion, U.S. Patent, Editor. The Procter & Gamble Company, USA
[115] E.J.-L. Perrier, A.R. Huc, Method for producing nanocapsules with crosslinked protein-based walls nanocapsules thereby obtained and cosmetic, pharmaceutical and food compositions using same, U.S. Patent, Editor. Coletica, USA, 1992.
[116] T.R. Tice, R.M. Gilley, Microencapsulation process and products therefrom, U.S. Patent, Editor. Southern Research Institute, USA, 1995
[117] S.A. Jenekhe, X.L. Chen, Macromolecular self-assembly of microstructures, nanostructures, objects and mesoporous solids W.I.P. Organization, Editor. University of Rochester, Switzerland, 1999.
[118] A. Baniel, Sweetener compositions W.I.P. Organization, Editor. Innova Sa Baniel, A., Switzerland, 2007.
[119] M. Maxfield, B.R. Christiani, Polymer nanocomposites formed by melt processing of a polymer and an exfoliated layered material derivatized with reactive organo silanes, W.I.P. Organization, Editor. Allied Signal Inc., Switzerland, 1993.
[120] M. Mawatari, C. Hamazaki, T. Furuyama, Antibacterial resin composition, U.S. Patent, Editor. Japan Synthetic Rubber Co., Ltd., USA, 1997.
[121] T.P.o.E. Nanotechnologies, Nanotechnology consumer products inventory, 2013. Available from: http://www.nutritionaloutlook.com/article/big-potential
[122] T.P.o.E. Nanotechnologies, Nanotechnology consumer products inventory, 2013.
[123] AquaNova, 2013. Available from: http://www.aquanova.de/media/public/pdf_produkteunkosher/NovaSOL_BCS.pdf

[124] Nanotechproject, 2011. Available from: http://www.nutralease.com/Nutra/Templates/showpage.asp?DBID=1&LNGID=1&TMID=84&FID=767
[125] T. International., 2006. Available from: http://www.taiyointernational.com/products/sunactive/
[126] Wild Flavors Inc., 2013. Available from: http://www.wildflavors.com/EMEA-EN/products/natural-flavor-ingredients/colors-from-nature/
[127] Life Enhancement, 2013. Available from: http://www.life-enhancement.com/shop/product/nsr-nanoresveratrol
[128] NanoSinergy, 2013. Available from: http://sprayforlife.com/
[129] Solgar, 2013. Available from: http://www.solgar.com/SolgarProducts/Nutri-Nano-CoQ-10-Alpha-Lipoic-Acid-Softgels.htm
[130] L. Laboratories, Altrient. Available from: http://www.livonlabs.com/europe.html
[131] B.N. Delivery, Bioral. Available from: http://www.bloomberg.com/research/stocks/private/snapshot.asp?privcapId=4690005
[132] Improveat, 2014. Available from: http://www.improveat.pt/en/products/
[133] T.P.o.E. Nanotechnologies, Nanotechnology consumer products inventory, 2007. Available from: http://www.nanotechproject.org/cpi/products/bluemoongoodstm-fresh-box-silver-nanoparticle-food-storage-containers/
[134] M.P. Corporation, N-Coat. Available from: http://multifilm.com/pouch-film/n-coat/
[135] T.P.o.E. Nanotechnologies, Nanotechnology consumer products inventory, 2007. Available from: http://www.nanotechproject.org/cpi/products/fresherlongertm-miracle-food-storage/
[136] T.P.o.E. Nanotechnologies, Nanotechnology consumer products inventory. Available from: http://www.nanotechproject.org/cpi/products/nano-silver-baby-mug-cup/
[137] Nanocor®, Polymer Polymer-Clay Nanocomposites Clay Nanocomposites-Better Plastics Better Plastics. Available from: http://www.epa.gov/oppt/nano/p2docs/casestudy2_lan.pdf
[138] I.N. Inc. Nansulate®, Available from: http://www.industrial-nanotech.com/main-page.html
[139] Bioni, Bioni's nano paint, 2008. Available from: http://www.bioni.de/en/index.php?page=news&lang=en-news040
[140] T.P.o.E. Nanotechnologies, Nanotechnology consumer products inventory, 2007. Available from: http://www.nanotechproject.org/cpi/products/nano-air-filter/
[141] C.E. Handford, M. Dean, M. Henchion, M. Spence, C.T. Elliott, K. Campbell, Implications of nanotechnology for the agri-food industry: opportunities, benefits and risks, Trends Food Sci. Technol. 40 (2) (2014) 226–241.
[142] V. Amenta, K. Aschberger, M. Arena, H. Bouwmeester, F. Botelho Moniz, P. Brandhoff, S. Gottardo, H.J.P. Marvin, A. Mech, L. Quiros Pesudo, H. Rauscher, R. Schoonjans, M.V. Vettori, S. Weigel, R.J. Peters, Regulatory aspects of nanotechnology in the agri/feed/food sector in EU and non-EU countries, Regul. Toxicol. Pharmacol. 73 (1) (2015) 463–476.
[143] E. Commission, The EU Framework Programme for Research and Innovation, 2014. Available from: http://ec.europa.eu/programmes/horizon2020/en/h2020-section/nanotechnologies
[144] USDA, USDA Awards $3.8 Million in Grants for Nanotechnology Research, 2015. Available from: http://nifa.usda.gov/sites/default/files/resource/AFRI Nanotech Release 2015 Fact Sheet P1.pdf
[145] J. Roosen, A. Bieberstein, S. Blanchemanche, E. Goddard, S. Marette, F. Vandermoere, Trust and willingness to pay for nanotechnology food, Food Policy 52 (2015) 75–83.
[146] M. Siegrist, M.-E. Cousin, H. Kastenholz, A. Wiek, Public acceptance of nanotechnology foods and food packaging: the influence of affect and trust, Appetite 49 (2) (2007) 459–466.
[147] G. Zhou, Nanotechnology in the Food System: Consumer Acceptance and Willingness to Pay, in Agricultural Economics. 2013, University of Kentucky.
[148] C.E. Handford, M. Dean, M. Spence, M. Henchion, C.T. Elliott, K. Campbell, Awareness and attitudes towards the emerging use of nanotechnology in the agri-food sector, Food Control 57 (2015) 24–34.

CHAPTER THREE

Small Solutions to Large Problems? Nanomaterials and Nanocomposites in Effluent, Water, and Land Management

Andrew B. Cundy
University of Southampton, Southampton, United Kingdom

Contents

3.1 INTRODUCTION

Nanoscience and nanotechnologies, involving research and technology development at the atomic, molecular, or macromolecular levels in the length scale of approximately 1–100 nm, are finding increasing application in the wider food industry—in agricultural production (e.g., nanocapsules for pesticide or vaccine delivery), food processing (e.g., as gelation or viscosifying agents), food packaging (e.g., as sensors or barrier materials), and supplements (e.g., as drug or neutraceutical delivery agents) ([1], various chapters in this volume). Indeed, the "nanofood" market (i.e., where nanotechnology techniques or tools are used during cultivation, production, processing, or packaging of food) is rapidly expanding, with an estimated global value of $410 million in 2006, predicted to grow to several billion dollars in the current decade [2]. In terms of the wider food production industry, nanotechnology has also been heralded as a potential solution to many key water purification, waste water and effluent treatment, and soil and groundwater management, issues (e.g., [3–5]) including those deriving from intensive agriculture. The supply of clean drinking water (and indeed water used for irrigation and food and drink production), and the effective treatment of waste and drainage waters to protect human health (and that of other environmental receptors) are key societal

Emerging Nanotechnologies in Food Science

challenges, particularly for a range of emerging contaminants which are increasingly becoming ubiquitous in the environment. These emerging contaminants include polar herbicides used in crop production, plasticizers (for instance produced in food packaging), endocrine disrupting compounds (EDCs), and pharmaceutical/animal medicine products, for which existing water and effluent treatment processes (e.g., ozonation, activated carbon) may not be effective in reaching regulatory standards, or in removing the contaminants of concern to residual concentrations that pose only a negligible threat to the environment (see review in Ref. [6]). This chapter discusses the current (and potential) applications of nanotechnology in the treatment of emerging contaminants and other agricultural and food production wastes and effluents, and outlines recent research on nanocomposites and nanostructured materials aimed at producing scalable, low-cost and nontoxic devices for effective effluent and water treatment and land remediation and regeneration.

3.2 NANOMATERIALS/NANOPARTICLES AS EFFLUENT, WASTEWATER, AND SOIL CLEAN-UP TOOLS

Some nanoparticles occur naturally and are ubiquitous in air, water, soils, and sediments (e.g., clay minerals have at least one dimension in the nanoscale, as do colloids, some aerosols, etc.), playing a key role in many atmospheric, oceanographic, geological, and biological cycles and processes (e.g., [7,8]). Nanoparticles may also, however, be produced specifically for a range of emerging scientific and engineering applications, as so-called engineered nanoparticles, which since the 1990s have been proposed for use in a variety of environmental applications, including water purification, waste water treatment, indoor and outdoor air cleaning, and soil and groundwater remediation [3]. The use of nanoparticles in environmental clean-up applications largely makes use of their enhanced reactivity, surface area, and/or enhanced mobility [9]. For example, zero valent (elemental) iron nanoparticles of 1–100 nm have an active surface area in the range of 20–40 m^2/g, providing 10–1000 times greater reactivity than granular iron, which has a surface area <1 m^2/g [10]. The small size of nanoparticles results in a high percentage of surface atoms (and surface energy), which can generate novel properties in relation to their bulk counterparts: for example, gold (and other noble metals) at bulk scale is unreactive, whereas at the nanoscale gold is highly reactive and has been utilized in nanoparticle form for the detection of pesticides [11] and for the catalytic destruction of recalcitrant halocarbons [12], while silver in particular has been used in a variety of antibacterial applications in textiles and other manufactured goods, and in water remediation [13]. It is also possible to surface modify or functionalize nanoparticles to target specific contaminants or reaction pathways [14]. Current environmental applications of nanoparticles use a wide variety of particle types: zerovalent iron, bimetallic alloys, iron oxide minerals, nanoscale zeolites, zinc oxides, carbon nanotubes and fibers, titanium and zirconium oxide minerals, gold and silver, etc. in both adsorptive and catalytic applications. Some of the nanoscale metal oxides (e.g., nanomaghemite and nanomagnetite) are superparamagnetic, which can facilitate their separation and recovery following application using a low-gradient magnetic field.

Nanotechnologies have been applied in a range of systems of relevance to food and agricultural production, to treat both concentrated and more dilute aqueous waste streams, to remove persistent or recalcitrant contaminants, and to clean-up contaminated soils and groundwaters. Rahman et al. [15] review the use of nanoparticle technologies for the treatment of anaerobic digestor residues and other relatively concentrated wastes in bioenergy generation, and note both inhibitory and enhancing effects on energy conversion depending on nanoparticle type and concentration, while the use of metal and carbon nanoparticles for catalytic destruction of trace organics, dyes, and other contaminants in waste waters has been an area of major research effort (e.g., [14,16–21]). For waste treatment and land remediation or clean-up applications, field trials and prototype devices have generally used nanofilters, bead-type devices, or directly injected nanoparticles as clean-up tools. For example, the ArsenX^{np} system [22] is a hybrid (regenerable) sorbent consisting of nanoparticles of hydrous iron oxide distributed through porous polymeric beads, which has been applied for arsenic removal from groundwater-sourced drinking water, while nanofiltration is a relatively well-established technique for water purification and desalination [23] and increasingly for food processing. Direct injection of nanoparticles for groundwater clean-up has made extensive use of nano zero-valent iron (NZVI) and bimetallic nanoparticles, particularly to target chlorinated solvent contamination [24–26]. The particles used are typically injected as a slurry directly into the subsurface environment to remediate contaminated groundwater plumes or contaminant source zones, thus avoiding the need for intrusive digging or pumping methods, and may be suspended in a hydrophobic fluid (i.e., injected as an emulsion) to prevent particle agglomeration or passivation (a term used to describe the oxidation or inactivation of iron nanoparticles) before reaching the contaminants they are intended to react with [9]. A number of authors have also developed and applied NZVI on colloidal carbon, mineral, or resin hosts for the reduction or removal of Cr, Pb, U, nitrate, dyes, and other effluent and groundwater contaminants (e.g., [27–32]). For applications in topsoil nanoparticles can be mixed in using conventional agricultural practices (e.g., ploughing) although most practical applications have used direct underground injection (e.g., [3]), with application of bulk amendments such as iron grits or chemical solution additives being more commonly applied in topsoils for contaminant stabilization [9]. Pan et al. [33], however, report the successful immobilization of phosphorous in topsoils by adsorption using nanoscale magnetite (Fe_3O_4) particles, and there is potential to utilize nanoparticles and nanostructured materials for delivery or retention of soil nutrients and for general improvement of soil characteristics (e.g., [34]).

Although nanomaterials and nanoparticles have shown utility at bench and full/field scale for a range of applications in the food industry, in agriculture and in land, water and wastes management, serious concerns have been raised concerning the health implications of widespread nanoparticle use and release, deriving largely from the small size and high reactivity and potential mobility (in both environmental and biological systems) of engineered nanoparticles. For example, environmental exposure to nanoparticles may allow their penetration into the deep lung via inhalation, where the clearance mechanism may not eject highly anisotropic, nonbiodegradable nanomaterials [35]. Furthermore, the passage of nanoparticles across cell membranes directly into cells or tissues may interfere with important cellular functions, while their enhanced reactivity may also have free radical-releasing, proinflammatory properties ([36], see also Chapter 5 of this volume). The potential for trophic transfer of nanoparticles (i.e., movement through the food chain) has also been identified, particularly in food crops (e.g., [37–40]). In addition, there are serious cost issues in the bulk use of many novel nanomaterials, particularly where solid or aqueous treatment volumes are large (e.g., in bulk wastewater treatment, or management of agricultural solids, soils and residues) or where applications have low added economic value. Questions over the scalability of many treatment processes remain unresolved, for example, the need to recover and recycle dispersed nanoparticles from bulk treatment systems (e.g., wastewater treatment plants), and how to avoid significant back pressures in nanofiltration devices when filtering large volumes.

The use of nanoparticles in a static, or contained, system however may avoid the potential health and environmental problems caused by their application as "free" nanoparticles, while also giving more flexibility to their range of potential applications in effluent, water and soil treatment or amendment. Indeed, the Royal Society's (UK) key 2004 assessment on "Nanoscience and Nanotechnologies: Opportunities and Uncertainties" [36] notes that:

> *Currently, we see health, safety and environmental hazards of nanotechnologies as being restricted to discrete manufactured nanoparticles and nanotubes in a free rather than embedded form.*

This "nanocomposite" approach (Fig. 3.1) involves physically or chemically embedding the nanoparticles into a bulk carrier, often in bead, fibrous, or column/monolithic format, hence retaining the bulk of their reactivity while avoiding release to the surrounding environment. Alternatively, nanofunctionality can be generated via use of tailored nanostructured bulk materials or nanoassembled devices (e.g., Whitby et al. [41] report use of assembled multiwalled carbon nanotubes, or buckycolumns, as controllable filtration devices). This chapter presents recent work carried out at the University of Brighton and with European academic and industrial partners which examines the use of nanocomposite and nanostructured devices as high through-flow, flow-over or bead-type reactive devices for treatment of liquid wastes, water and soils, and for wider

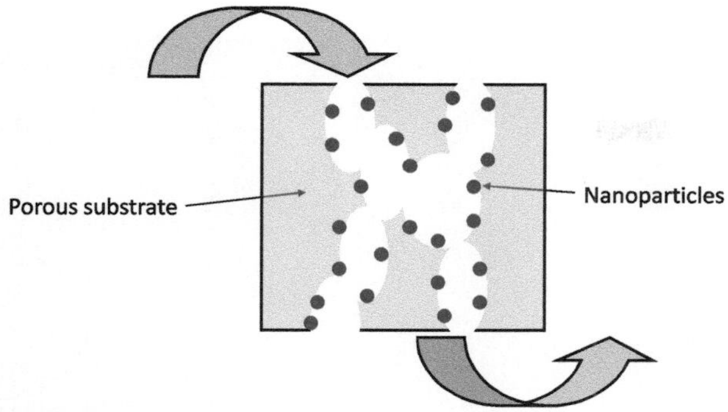

Figure 3.1 ***Schematic of nanocomposite column filter.*** Reactive nanoparticles are physically or chemically embedded into the surface of a low-cost macroporous scaffold, and react with contaminated water flowing through the filter.

separation applications.The overall aim of the work programs described here is to produce low-cost and scalable technologies for the effective treatment of groundwaters, surface waters, soils and liquid effluents and wastes (including agricultural wastes) contaminated with heavy metals, metalloids (e.g., arsenic), nutrients, and organic contaminants such as EDCs, pharmaceutical/animal medicine products, and pesticides/herbicides.

3.2.1 Case 1: Cryogel-based nanocomposites

Cryogels are macroporous gel materials produced by the polymerization of water-soluble monomers under freezing temperatures (so-called cryogelation). Ice crystals in the frozen material act as porogens (i.e., pore formers), and are washed out following defrosting to leave a macroporous gel structure. Cryogels are easy to manufacture (commonly requiring only one cycle of freezing–defrosting of the reagent/polymer solution), nontoxic, and offer controllable permeability, high mechanical strength, chemical stability, and shape recovery [42,43]. Their macroporous structure (containing large interconnected pores of up to 100 μm diameter) gives low flow resistance and low mass transfer limitations (and so low back pressures), and they can be produced in a variety of geometries [44]. Cryogels have been utilized as effective hosts or porous scaffolds for nanocomposite filtration devices targetting a range of problem contaminants. For example, Savina et al. [44] used hematite and magnetite nanoparticles embedded in the thin polymer walls surrounding the macropores of HEMA-based cryogels as adsorbents (Fig. 3.2A) to remove As(III) (a major global groundwater and drinking water contaminant) from simulated environmental waters, and showed that despite physical embedding of the nanoparticles into the polymer high nanoparticle reactivity was retained due to short diffusion pathways. Performance data indicated rapid and effective sorption of As(III) across a wide pH range (3–9), even in the presence of competing ions such as phosphate. Önnby et al. [45] targeted the same contaminant (but in its higher oxidation state) using aluminum

Figure 3.2 (A) SEM image of cryogel-Fe_3O_4 composite, showing interconnected macroporous structure of composite material. Fe_3O_4 nanoparticles are embedded in the thin polymer walls surrounding the macropores. (B) Large-scale cryogel produced by the prefreezing method outlined in Savina et al., [50], which allows production of large-volume cryogels with reproducible control of the macroporous structure throughout the sample, and an even distribution of particles within the 3D structure of the nanocomposite. *[Image (with permission) from I.N. Savina, C.J. English, R.L.D. Whitby, Y. Zheng, A. Leistner, S.V. Mikhalovsky, A.B. Cundy, High efficiency removal of dissolved As (III) using iron nanoparticle-embedded macroporous polymer composites, J. Hazard. Mater. 192 (2011) 1002–1008 [44]].*

oxide nanoparticles embedded into polyacrylamide-based cryogels, and showed effective removal of As(V) from test waters via adsorption. Any washout or leaching of aluminum oxide nanoparticles was undetectable during relatively aggressive filtration conditions, indicating the effectiveness of the embedding process, while in vitro cytotoxicity studies using heterogeneous human epithelial colorectal adenocarcinoma cells showed no toxicity of leachates after initial washing of the nanocomposites to remove unreacted monomer. Flow rates during treatment by these cryogel-based nanocomposites exceed those of conventional bead-type devices, and the embedding of the nanoparticles prevents their agglomeration and further loss of reactivity, which may be a problem in some applications involving free release of nanoparticles (e.g., [24,46]). Use of such macroporous polymers has also been attempted successfully at pilot scale in moving bed reactors for waste water treatment [47]. Practically, freezing kinetics during conventional cryogelation methods limit the size of cryogel-based composites to a few milliliters in volume, or with at least one small (ca. 2 cm or less) dimension (e.g., [48,49])—larger samples have variable pore size distribution across the cryogel matrix due to differential freezing rates across the sample. In addition, slower freezing in larger samples allows nanoparticle separation (by floating, sinking, or agglomeration) during the freezing process, which results in the failure of the composite entirely or the formation of macroporous gels with uneven particle distribution. Savina et al. [50], however, report a partial or prefreezing method which controls heat transfer during cryogelation and allows the production of large volume (400 mL or greater) cryogel-based nanocomposites, with effective and reproducible control of the macroporous structure throughout the sample, and an even distribution of particles (including those which do not easily form a stable suspension) within the 3D structure of the nanocomposite (Fig. 3.2B). This method allows reproducible production of large volume (i.e., scalable) nanocomposites for a range of biomedical, biotechnological, bioseparation, and environmental applications.

3.2.2 Case 2: Silica-silver nanocomposites

A number of authors have proposed the use of mesoporous silica as a low-cost scaffold or host for reactive nanoparticles, or for producing nanocomposites for applications such as medical imaging and drug delivery (e.g., [51–54]). Katok et al. [55] report the chemical generation of silicon hydride groups (—SiH) on low-cost mesoporous silica substrates (via washing with triethoxysilane in the presence of acetic acid), which are in turn used (via reduction of silver nitrate solution) for the controlled deposition of discrete silver nanoparticles onto the silica surface (Fig. 3.3A). This process essentially produces immobilized (and size controlled) silver nanoparticles on a low-cost and adaptable substrate, which can then be applied to effluent, waste and water treatment, either as a filtration device or as a reactive surface. Silver nanoparticles are well known, and have been extensively utilized, for their antibacterial properties (see also Chapter 6 of this volume), and nanosilver is also capable of highly effectively removing pesticide residues and other organic contaminants [13]. In addition, nanosilver very effectively reacts

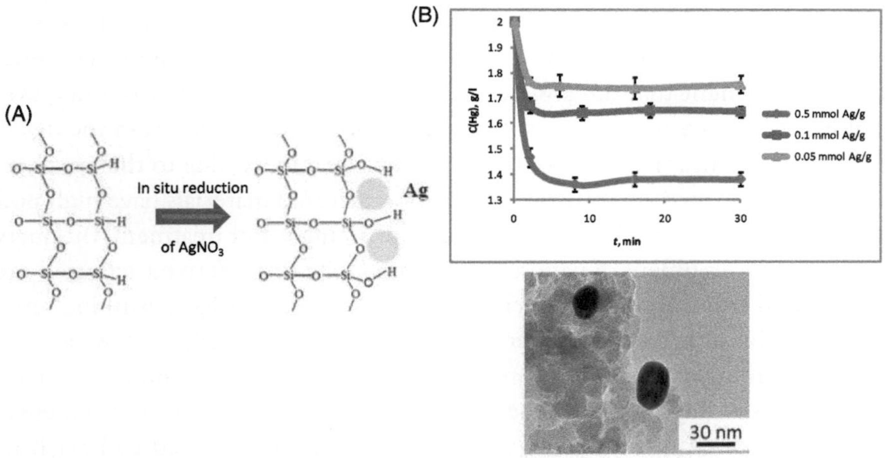

Figure 3.3 (A) Schematic of size-controlled Ag nanoparticle generation on triethoxysilane-modified mesoporous silica scaffold. (B) Rapid removal of mercury via Ag:Hg amalgam formation. Kinetic data are shown for three different Ag NP "loadings," for a 312 mg/L mercury stock solution. Lower photograph shows TEM image of silver:mercury amalgam nanoparticles on the modified silica surface. *(Adapted from K.V. Katok, R.L.D. Whitby, T. Fukuda, T. Maekawa, I. Bezverkhyy, S.V. Mikhalovsky, A.B. Cundy, Hyperstoichiometric interaction between silver and mercury redefines conventional redox chemistry at the nanoscale, Angew. Chem. Int. Ed. 51 (2012) 2632–2635 [55]).*

with the toxic heavy metal mercury, readily forming an amalgam (a similar but inverse process has been used extensively in silver and gold mining, whereby mercury amalgamation is used for precious metal extraction, e.g., [56,57]). In laboratory-based trials, size-controlled silver nanoparticles (of 11–45 nm size) were generated on a triethoxysilane-modified silica surface via reduction of silver cations, and were shown to rapidly and effectively remove mercury from aqueous solution via formation of a Schachnerite amalgam ($Ag_{1.1}Hg_{0.9}$ [55] with 1 g of Ag capable of treating > 10,000 m^3 of contaminated water (at environmentally realistic ppb levels of Hg(II)). Reaction kinetics were extremely rapid (<5 min, or effectively instantaneous, Fig. 3.3B) and the authors report an unusually high mercury-to-silver stoichiometric ratio when silver nanoparticle size is below 32 nm—effectively generating greatly enhanced removal of mercury from the treated solutions below a critical nanoparticle size. The low-cost surface modification technique (and low cost of the silica raw material) used lends itself to producing reactive surfaces in a variety of geometries, which can then be integrated into larger devices for effluent and water treatment for mercury, pesticide residues, microbial contaminants, and other target substances.

3.2.3 Case 3: Nanostructured tailored activated carbons

Nanostructured materials (i.e., "tailored" materials with controlled nanoporosity or nanoscale functionality) also show considerable potential in targeted contaminant removal from wastes, effluents, and treated waters, or in separation of active biological molecules. Nanostructured materials or devices present the enhanced properties typical of nanomaterials, such as high surface area in relation to their volume and enhanced reactivity, with the added benefit that the absence of nanoparticles makes their use free from the potential hazards associated with possible leaching of nanoparticles from the devices, or physico-chemical transformations of the nanoparticles. Hence, due to their relative safety and high performance and functionality, nanostructured materials have high potential for direct application with food products and in drinking water treatment. Busquets et al. [58] report the development of nanostructured phenolic-resin derived activated carbons, used for targeted removal of the pesticide metaldehyde from surface, drinking, and waste waters. Metaldehyde is an 8-member cyclic tetramer of acetaldehyde widely used as a molluscicide in agriculture and domestic gardening to control slugs and snails. This polar molecule is highly mobile in the aqueous environment, is not removed effectively in conventional wastewater treatment plants, and has been observed to breach regulatory limits (of 0.1 μg/L, set by the Drinking Water Directive, Council of the European Communities [59]) in surface waters and some drinking waters following heavy rainfall and wash-off from agricultural land. Attrition resistant activated carbon beads with controlled pore distributions, surface functionalization, and activation degree were synthesized from porous phenolic resin (following the technology patented by Tennison et al. [60]), and tested for metaldehyde adsorption. Adsorption of metaldehyde was, unusually, independent of specific surface area (S_{BET}) but was strongly favored in carbons with an absence of negatively charged functional groups, and with high microporosity (<2 nm) (with narrow pore size distribution) coupled with mesopores (2–50 nm) to allow effective diffusive transport. The maximum adsorption capacity for metaldehyde was 76 mg metaldehyde/g of carbon, compared to 13 mg metaldehyde/g carbon in conventional (i.e., standard industrial) granular activated carbon (GAC, Fig. 3.4A). Adsorption isotherms indicate strong adsorption of metaldehyde onto the phenolic carbon beads at environmentally realistic concentrations, and relatively rapid adsorption kinetics were achieved. Adsorption efficiency was maintained even in the presence of high background salts and organic matter, which is likely to be an effect of the porous "skin" of the carbons limiting the diffusion of large molecules, and possibly the presence of metaldehyde-selective sites on the carbons. The high attrition resistance of the beads makes them appropriate for a range of water (and gas) treatment applications (e.g., column filtration, moving bed applications, membrane reactors, e.g., [61]), and their manufacturing method is easily scalable to industrial scale production, although they are significantly more expensive than conventional (industry-standard) GAC. The higher cost can to some extent be offset by the potential to steam regenerate the carbons on-site rather than using high temperature off-site regeneration. Additional functionality can be added to the beads through incorporation of metallic nanoparticles into their porous structure, either by adding metal salts to the phenolic resin precursors prior to mixing and activation, or by precipitation onto the carbon surface following activation (Fig. 3.4B). The utility of these nanometal–carbon composites for treatment of mixed

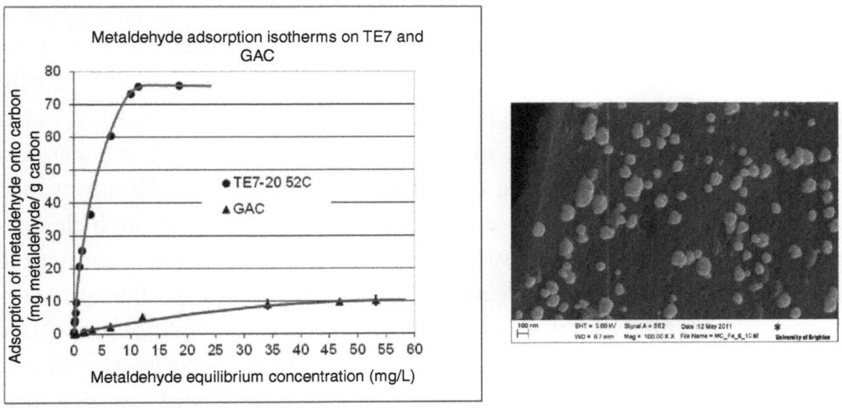

Figure 3.4 (A) Adsorption isotherms for the removal of metaldehyde by the nanostructured phenolic-resin derived activated carbon TE7-20-52C, and conventional granular activated carbon (GAC). (B) SEM backscatter image of nanostructured carbon with additional functionality introduced by precipitation of Fe/Cu nanoparticles on the carbon surface. *(From data presented in R. Busquets, O.P. Kozynchenko, R.L.D. Whitby, S.R. Tennison, A.B. Cundy, Phenolic carbon tailored for the removal of polar organic contaminants from water: a solution to the metaldehyde problem? Water Res. 61 (2014) 46–56 [58]).*

contaminant streams (i.e., effluents and waters containing a combination of organic, inorganic, and in some cases radioactive contaminants) is currently being examined in the EU Framework 7 Industry Academia Partnerships and Pathways project WaSClean (Water and Soil Clean-Up from mixed contaminants, http://www.saske.sk/wasclean/), with particular focus on the application of Fe/Cu–carbon composites as column or bed filters, and as reactive-barrier type materials for groundwater treatment. These "tailored" carbon beads can also be effectively combined with cryogel technologies (as discussed in case 1), to produce hierarchical column filtration devices (i.e., with nano/micro, meso, and macro-porosity) for biological separation and contaminant removal applications [e.g., Busquets et al. [62] who show improved pore diffusion and effective removal of the pesticides atrazine and malathion within these hierarchical carbon:cryogel devices, using water with high total organic carbon (TOC) levels and high salinity].

3.3 DISCUSSION—TOWARDS PRACTICAL APPLICATION

The aforementioned examples show the utility of combining advances in materials science with nanoparticle technologies to produce flexible nanocomposite devices for water, waste, and effluent treatment and other environmental applications, which maintain the high reactivity of the nanoparticles while preventing their release to the surrounding environment. Nanoporous or nanostructured devices, such as those outlined in case 3, may also form highly effective contaminant removal or recycling devices. Although many current embodiments of nanocomposite devices are in bead or fibrous format, use of flexible, low-cost scaffolds such as modified silica surfaces, polymers, etc. allows a variety of device configurations and structures to be developed, targeted at particular end-use applications. For example, devices may be applied as high throughflow filters, membranes, wafers, reactive barrier materials, or as bed reactors or pipe-lining devices. Many nanocomposites and nanostructured materials are readily scalable and/or mass producible, for example, as in case 1, where liter volume cryogel composites can be produced which are capable of effective separation and filtration applications at large treatment volumes (Savina et al. [50]). The flexibility of such devices means that they can be integrated into existing waste, water, and effluent treatment and land remediation systems, at point of treatment or as decentralized solutions, rather than requiring additional infrastructure to be constructed. Use of nanocomposite systems may have a number of significant advantages over conventional waste, effluent, water, and land treatment systems in terms of their higher reactivity and capacity, more effective contaminant removal (including at low, environmentally relevant concentrations), lower waste production, reduced energy demand and carbon footprint, ability to be "retrofitted" to existing treatment facilities, and targeting of currently problematic contaminants. In addition, the flexible chemistry of the devices allows chemical fine-tuning to target specific ranges of contaminants, for example, pesticides in natural and treated waters, phthalates, EDCs, etc. A number of issues remain, however, in terms of the scale-up and adoption of these and of similar nanocomposite devices, particularly that of cost for their large-scale use, and regulatory and public acceptance/societal issues (e.g., [46]). For the former, effective device recyclability is key to reduce implementation and operational costs. For the latter, there remains in many parts of the world significant public opposition to the use of nanotechnology due to health concerns (despite the large numbers of market products such as textiles, sun creams, etc. that already contain nanoparticles), while implementing new technologies in water and land management remains problematic more generally, as noted previously by Cundy et al. [9], due to inherent conservatism in these sectors. Given the wide publicity over the potential health impacts of nanoparticles, this is a particularly severe constraint on the application of nanoparticle-based materials in drinking/potable water treatment, or indeed for direct application with food and drink products (although as noted in case 3, nanostructured materials have higher potential for direct application in these areas due to their relative safety and high performance and functionality). Treatment of industrial and sewage effluents, both to recycle potentially valuable waste products such as phosphate and rare elements and to remove problem contaminants such as estrogens and persistent pesticides, may be an easier route to upscaling and commercial application for nanoparticle-containing composites.

As noted by Hillie and Hlolphe [63] 10 years ago, nanoenabled technologies for water treatment are already on the market, with nanofiltration currently the most mature technology, while a range of nanobased technologies are applied in the food industry (this volume). Although a number of barriers remain for the large-scale application of nanoenabled technologies in the food and environmental sectors, not least cost, upscaling and regulatory considerations, the significant improvements in waste, effluent, water, and soil treatment offered by these technologies coupled with the increasing use of nanotechnologies in other sectors mean that their adoption and development is likely to increase significantly in the future.

ACKNOWLEDGMENTS

The author acknowledges the support of the following funding sources: the CommercialiSE POC fund (SEEDA, ESF), Marie Curie FP7 Intra-European and International Incoming Fellowships programme, Leverhulme Trust, FP7 IAPP programme (Carbosorb, proj no. 230676, WasClean, proj no. 612250), and industry and academic partners within the projects discussed. This paper is based on presentations made at the 2011 International Symposium on Environmental Science and Technology (Dongguan, China), and the 2015 WasClean project mid-term meeting (Thessaloniki, Greece).

REFERENCES

[1] Nanowerk, Food nanotechnology. 2015. www.nanowerk.com/nanotechnology-in-food.php.

[2] T. Joseph, M. Morrison, Nanotechnology in agriculture and food, Nanoforum report, 2006. Available from: ftp://ftp.cordis.europa.eu/pub/nanotechnology/docs/nanotechnology_in_agriculture_and_food.pdf

[3] N.C. Mueller, B. Nowack, Nanoparticles for remediation: solving big problems with little particles, Elements 6 (2010) 395–400.
[4] S. Ragan, R.L.D. Whitby, A.B. Cundy, Old and new carbons for wastewater treatment, Water Sewerage J. 3 (2012) 34–35.
[5] X. Qu, P.J.J. Alvarez, Q. Li, Applications of nanotechnology in water and wastewater treatment, Water Res. 47 (2013) 3931–3946.
[6] K.E. Murray, S.M. Thomas, A.A. Bodour, Prioritizing research for trace pollutants and emerging contaminants in the freshwater environment, Environ. Pollut. 158 (2010) 3462–3471.
[7] M.F. Hochella, Nanoscience and technology: the next revolution in the earth sciences, Earth Planet. Sci. Lett. 203 (2002) 593–605.
[8] M.F. Hochella, D. Aruguete, B. Kim, A.S. Madden, Naturally occurring inorganic nanoparticles: general assessment and a global budget for one of Earth's last unexplored geochemical components, in: G. Haibo, B. Amanda (Eds.), Nature's Nano-Structures, first ed., Pan Stanford Publishing, Victoria, Australia, 2012.
[9] A.B. Cundy, L. Hopkinson, R.L.D. Whitby, Use of iron-based technologies in contaminated land and groundwater remediation: a review, Sci. Total Environ. 400 (2008) 42–51.
[10] C.B. Wang, W.X. Zhang, Synthesizing nanoscale iron particles for rapid and complete dechlorination of TCE and PCBs, Environ. Sci. Technol. 31 (1997) 2154–2156.
[11] K.P. Lisha, T. Pradeep, Enhanced visual detection of pesticides using gold nanoparticles, J. Environ. Sci. Health B 44 (2009) 697–705.
[12] A.S. Nair, T. Pradeep, Halocarbon mineralization and catalytic destruction by metal nanoparticles, Curr. Sci. 84 (2003) 1560–1564.
[13] T. Pradeep, Noble metal nanoparticles for water remediation: a critical review, Thin Solid Films 517 (2009) 6441–6478.
[14] K.V. Voitko, R.L.D. Whitby, V.M. Gun'ko, O.M. Bakalinska, M.T. Kartel, K. Laszlo, A.B. Cundy, S.V. Mikhalovsky, Morphological and chemical features of nano- and macroscale carbons affecting hydrogen peroxide decomposition in aqueous media, J. Colloid Interface Sci. 361 (2011) 129–136.
[15] K. Rahman, L. Melville, I. Huq, S. Khoda, Understanding bioenergy production and optimisation at the nanoscale—a review, J. Exp. Nanosci. 11 (2016) 762–775.
[16] S. Shylesh, V. Schünemann, W.R. Thiel, Magnetically separable nanocatalysts: bridges between homogeneous and heterogeneous catalysis, Angew. Chem. Int. Ed. 49 (2010) 3428–3459.
[17] M. Zahmakiran, S. Özkar, Metal nanoparticles in liquid phase catalysis; from recent advances to future goals, Nanoscale 3 (2011) 3462–3481.
[18] Y.C. Zhang, J. Li, M. Zhang, D.D. Dionysiou, Size-tunable hydrothermal synthesis of SnS_2 nanocrystals with high performance in visible light-driven photocatalytic reduction of aqueous Cr(VI), Environ. Sci. Technol. 45 (2011) 9324–9331.
[19] R.M. Mohamed, D.L. McKinney, W.M. Sigmund, Enhanced nanocatalysts, Mater. Sci. Eng.: R 73 (2012) 1–13.
[20] K. Pathakoti, S. Morrow, C. Han, M. Pelaez, X. He, D.D. Dionysiou, H.-M. Hwang, Photoinactivation of *Escherichia coli* by sulfur-doped and nitrogen–fluorine-codoped TiO_2 nanoparticles under solar simulated light and visible light irradiation, Environ. Sci. Technol. 47 (2013) 9988–9996.
[21] Y. Wang, J. Liu, P. Wang, C.J. Werth, T.J. Strathmann, Palladium nanoparticles encapsulated in core–shell silica: a structured hydrogenation catalyst with enhanced activity for reduction of oxyanion water pollutants, ACS Catal. 4 (2014) 3551–3559.
[22] P. Sylvester, P. Westerhoff, T. Moller, M. Badruzzaman, O. Boyd, A hybrid sorbent utilising nanoparticles of hydrous iron oxide for arsenic removal from drinking water, Environ. Eng. Sci. 24 (2007) 104–112.
[23] A.W. Mohammed, Y.H. Teow, W.L. Ang, Y.T. Chung, D.L. Oatley-Radcliffe, N. Hilal, Nanofiltration membranes review: recent advances and future prospects, Desalination 356 (2015) 226–254.
[24] R.P. Bardos, B. Bone, D. Elliott, N. Hartog, J. Henstock, P. Nathanail, A risk/benefit approach to the application of iron nanoparticles for the remediation of contaminated sites in the environment. Defra Research Project Final Report CB0440 (2011).
[25] C. Su, R.W. Puls, T.A. Krug, M.T. Watling, S.K. O'Hara, J.W. Quinn, N.E. Ruiz, A two and half-year-performance evaluation of a field test on treatment of source zone tetrachloroethene and its chlorinated daughter products using emulsified zero valent iron nanoparticles, Water Res. 46 (2012) 5071–5084.

[26] C. Su, R.W. Puls, T.A. Krug, M.T. Watling, S.K. O'Hara, J.W. Quinn, N.E. Ruiz, Travel distance and transformation of injected emulsified zero valent iron nanoparticles in the subsurface during two and half years, Water Res. 47 (2013) 4095–4106.

[27] S. Ponder, J. Darab, T. Mallouk, Remediation of Cr(VI) and Pb(II) aqueous solutions using supported, nanoscale zero-valent iron, Environ. Sci. Technol. 34 (2000) 2564–2569.

[28] H. Park, Y.M. Park, K.M. Yoo, S.H. Lee, Reduction of nitrate by resin-supported nanoscale zero-valent iron, Water Sci. Technol. 59 (2009) 2153–2157.

[29] X. Fei, L. Cao, L. Zhou, Y. Gu, X. Wang, Degradation of bromamine acid by nanoscale zero-valent iron (nZVI) supported on sepiolite, Water Sci. Technol. 66 (2012) 2539–2545.

[30] Z. Chen, T. Wang, X. Jin, Z. Chen, M. Megharaj, R. Naidu, Multifunctional kaolinite-supported nanoscale zero-valent iron used for the adsorption and degradation of crystal violet in aqueous solution, J. Colloid Interface Sci. 398 (2013) 59–66.

[31] R.A. Crane, T. Scott, The removal of uranium onto carbon-supported nanoscale zero-valent iron particles, J. Nanopart. Res. 16 (2014) 2813.

[32] J. Busch, T. Meißner, A. Potthoff, S. Bleyl, A. Georgi, K. Mackenzie, R. Trabitzsch, U. Werban, S.E. Oswald, A field investigation on transport of carbon-supported nanoscale zero-valent iron (nZVI) in groundwater, J. Contam. Hydrol. 181 (2015) 59–68.

[33] G. Pan, L. Li, D. Zhao, H. Chen, Immobilisation of non-point phosphorous using stabilised magnetite nanoparticles with enhanced transportability and reactivity in soils, Environ. Pollut. 158 (2010) 35–40.

[34] R. Liu, R. Lal, Synthetic apatite nanoparticles as a phosphorous fertilizer for soy bean (Glycine Max.), Sci. Rep. 4 (2014) 5686.

[35] C.W. Lam, J.T. James, R. McLusky, R.L. Hunter, Pulmonary toxicity of single-wall carbon nanotubes in mice 7 and 90 days after intratracheal instillation, Toxicol. Sci. 77 (2004) 126–134.

[36] Royal Society, Royal Academy of Engineering, Nanoscience and nanotechnologies: opportunities and uncertainties, 2004. Available from: http://www.nanotec.org.uk/index.htm

[37] J. Hong, J.R. Peralta-Videa, C. Rico, S. Sahi, M.N. Viveros, J. Bartonjo, L. Zhao, J. Gardea-Torresdey, Evidence of translocation and physiological impacts of foliar applied CeO_2 nanoparticles on cucumber (Cucumis sativus) plants, Environ. Sci. Technol. 48 (2014) 4376–4385.

[38] J. Trujillo-Reyes, S. Majumdar, C.E. Botez, J.R. Peralta-Videa, J.L. Gardea-Torresdey, Exposure studies of core-shell Fe/Fe3O4 and Cu/CuO NPs to lettuce (Lactuca sativa) plants: are they a potential physiological and nutritional hazard?, J. Hazard. Mater. 267 (2014) 255–263.

[39] J. Hawthorne, R. De la Torre Roche, B. Xing, L.A. Newman, X. Ma, S. Majumdar, J. Gardea-Torresdey, J.C. White, Particle-size dependent accumulation and trophic transfer of cerium oxide through a terrestrial food chain, Environ. Sci. Technol. 48 (2014) 13102–13109.

[40] S. Majumdar, J. Trujillo-Reyes, J.A. Hernandez-Viezcas, J.C. White, J.R. Peralta-Videa, J.L. Gardea-Torresdey, Cerium biomagnification in a terrestrial food chain: influence of particle size and growth stage, Environ. Sci. Technol. 50 (2016) 6782–6792.

[41] R.L.D. Whitby, T. Fukuda, T. Maekawa, S.V. Mikhalovsky, A.B. Cundy, Real-time imaging of complex nanoscale mechanical responses of carbon nanotubes in highly compressible porous monoliths, Nanotechnology 21 (2010) 075707.

[42] I.N. Savina, V.M. Gun'ko, V.V. Turov, M. Dainiak, G.J. Phillips, I. Galaev, S.V. Mikhalovsky, Porous structure and water state in cross-linked polymer and protein cryo-hydrogels, Soft Matter 7 (2011) 4276–4283.

[43] B. Gyarmati, E.Z. Mészár, L. Kiss, M.A. Deli, K. László, A. Szilágyi, Supermacroporous chemically cross-linked poly(aspartic acid) hydrogels, Acta Biomater. 22 (2015) 32–38.

[44] I.N. Savina, C.J. English, R.L.D. Whitby, Y. Zheng, A. Leistner, S.V. Mikhalovsky, A.B. Cundy, High efficiency removal of dissolved As (III) using iron nanoparticle-embedded macroporous polymer composites, J. Hazard. Mater. 192 (2011) 1002–1008.

[45] L. Önnby, C. Svensson, L. Mbundi, R. Busquets, A.B. Cundy, H. Kirsebom, γ-Al_2O_3-based nanocomposite adsorbents for arsenic(V) removal: assessing performance, toxicity and particle leakage, Sci. Total Environ. 473-474 (2014) 207–214.

[46] B. Karn, T. Kuiken, M. Otto, Nanotechnology and in situ remediation: a review of the benefits and potential risks, Environ. Health Perspect. 117 (2009) 1823–1831.

[47] M. Le Noir, F.M. Plieva, B. Mattiasson, Removal of endocrine-disrupting compounds from water using macroporous molecularly imprinted cryogels in a moving-bed reactor, J. Sep. Sci. 32 (2009) 1471–1479.
[48] F.M. Plieva, M. Karl, M.R. Aguilar, D. Gomez, S.V. Mikhalovsky, I.Y. Galaev, B. Mattiason, Pore structure of macroporous monolithic cryogels prepared from poly(vinyl alcohol), J. Appl. Polym. Sci. 100 (2006) 1057–1066.
[49] V.M. Gun'ko, I.N. Savina, S.V. Mikhalovsky, Cryogels: morphological, structural and adsorption characterisation, Adv. Colloid Interface Sci. 187 (2013) 1–46.
[50] I.N. Savina, G.C. Ingavle, A.B. Cundy, S.V. Mikhalovsky, A simple method for the production of large volume 3D macroporous hydrogels for advanced biotechnological, medical and environmental applications, Sci. Rep. 6 (2016) 21154.
[51] L. Wei, N. Hu, Y. Zhang, Synthesis of polymer—mesoporous silica nanocomposites, Materials 3 (2010) 4066–4079.
[52] B. Naik, V. Desai, M. Kowshik, V.S. Prasad, G.F. Fernando, N.N. Ghosh, Synthesis of Ag/AgCl–mesoporous silica nanocomposites using a simple aqueous solution-based chemical method and a study of their antibacterial activity on *E. Coli*, Particuology 9 (2011) 243–247.
[53] J.E. Lee, N. Lee, T. Kim, J. Kim, T. Hyeon, Multifunctional mesoporous silica nanocomposite nanoparticles for theranostic applications, Acc. Chem. Res. 44 (2011) 893–902.
[54] E. Petala, K. Dimos, A. Douvalis, T. Bakas, J. Tucek, R. Zboril, M.A. Karakassides, Nanoscale zero-valent iron supported on mesoporous silica: characterization and reactivity for Cr(VI) removal from aqueous solution, J. Hazard. Mater. 261 (2013) 295–306.
[55] K.V. Katok, R.L.D. Whitby, T. Fukuda, T. Maekawa, I. Bezverkhyy, S.V. Mikhalovsky, A.B. Cundy, Hyperstoichiometric interaction between silver and mercury redefines conventional redox chemistry at the nanoscale, Angew. Chem. Int. Ed. 51 (2012) 2632–2635.
[56] J.O. Nriagu, Mercury pollution from the past mining of gold and silver in the Americas, Sci. Total Environ. 149 (1994) 167–181.
[57] L. Lacerda, W Salomons (Eds.), Mercury from Gold and Silver Mining: A Chemical Time Bomb?, Springer, Berlin, 1998, pp. 147.
[58] R. Busquets, O.P. Kozynchenko, R.L.D. Whitby, S.R. Tennison, A.B. Cundy, Phenolic carbon tailored for the removal of polar organic contaminants from water: a solution to the metaldehyde problem?, Water Res. 61 (2014) 46–56.
[59] Council of the European Communities, Directive of the European Parliament and of the Council on the Quality of Water Intended for Human Consumption (1998) 98/83/EC.
[60] S.R. Tenninson, O.P. Kozynchenko, V.S. Volodymyer, J. Blackburn, Porous Carbons, WO2001GB03560 20010807, US2011097583 (2000).
[61] A.A. Lapkin, S.R. Tennison, W.J. Thomas, A porous carbon membrane reactor for the homogenous catalytic hydration of propene, Chem. Eng. Sci. 13 (2002) 2357–2369.
[62] R. Busquets, A.E. Ivanov, L. Mbundi, S. Hörberg, O.P. Kozynchenko, P.J. Cragg, I.N. Savina, R.L.D. Whitby, S.V. Mikhalovsky, S.R. Tennison, H. Jungvid, A.B. Cundy, Carbon-cryogel hierarchical composites as effective and scalable filters for removal of trace organic pollutants from water, J. Environ. Manage. 182 (2016) 141–148.
[63] T. Hillie, M. Hlophe, Nanotechnology and the challenge of clean water, Nat. Nanotechnol. 2 (2007) 663–664.

CHAPTER FOUR

Analysis of Nanomaterials in Food

Rosa Busquets
Kingston University London, Kingston upon Thames, United Kingdom

Contents

4.1 INTRODUCTION

The benefits that nanotechnology brings to the food industry are well established and will continue to rapidly evolve. Although there is a sharp increase in the number of food items that contain nanomaterials [1], research into the merger of nano- and food technologies is still in its infancy and still exploratory. As a result, the safety implications of the increasing number of food items containing nanotechnology entering the market remain unclear. In addition, it is well accepted that the rise in the production of items containing nanomaterials will lead to greater release of nanomaterials into the environment [2]. As such, there is pressing need for methodology to enable the study of nanomaterials in different media [3]. The terms nanomaterial and nanoparticle have been contrasted in Chapter 1. Here, the more general term between the two, nanomaterial, will be used when possible.

The development of products with nanotechnology in the food sector involves broad interdisciplinary expertise. The formulation of nanoingredients is studied by food scientists, who collaborate with material scientists to develop nanoenabled packaging materials with improved properties. The development of sensors based on nanotechnology to monitor food quality, the characterization of particles during the development of products, and the study of their potential release is carried out within the areas of analytical chemistry and material science. The safety of the nanomaterials used is examined in vitro, in vivo, and in the environment mainly by life scientists, who also assess nanomaterial–antibacterial properties where relevant. Consequently, synergy in these different areas of knowledge and approaches to understanding nanomaterials is needed to provide common understanding of the effects of nanomaterials in food and the environment and to help improve the efficacy and safety of the technology.

The early stages of the development of nanoingredients and products containing nanomaterials, such as inks, packaging materials, sensors, or water filters, usually involve a relatively high concentration of particles [4–6]. At this stage, the nanomaterials can be directly assessed by a range of detection systems with less extensive sample manipulation given that the sample matrix, if any, is not complex and the amount of particles available for the analysis is sufficient for instrument sensitivity during characterization. In contrast, when the nanomaterials are diluted within food, the environment, or biological tissues, the analytical procedures have to work at the limit of their detection capacity and incorporate extra steps such as preconcentration, clean-up and in some cases separation steps prior to detection. As such, the development of appropriate methodologies that can measure important properties of the particles relevant to their function within sample matrices with enough sensitivity is paramount. In addition, the ability to use a range of other complementary techniques is important. An example of a multiangled approach carried out for the characterization of a material containing nanoparticles was conducted by authors of this volume where they assessed the potential leakage of aluminum nanoparticles from composite water filters [5]. The composite water filters were rinsed, and the filtrate was passed through track-etched polycarbonate membranes (0.05 μm pore diameter). To measure possible particle leakage, the difference in mass of both the composite filter and polycarbonate membranes, before and after filtering water, was measured. The dried membranes were also examined with transmission electron microscopy (TEM), scanning electron microscopy (SEM), and particle-induced X-ray emission (PIXE). The latter was selected over SEM-EDX (also called EDS or energy dispersive X-ray) analysis because of its higher sensitivity. In addition, the safety of the leachates was assessed with cell toxicity assays and liquid chromatography-mass spectrometry. The cell tests selected (MTS assay) assessed the effect of the leachates on the integrity of the mitochondrial metabolic activity, a process vital in the survival of any cell [7], as well as the integrity of the cell membrane by assessing leakage of lactic dehydrogenase (LDH assay) that is released by dead cells [8]. Cell proliferation and toxicity assays have the advantage that cells, as *sensors* of toxicity, can be affected by a broad number of compounds or particles leached, and therefore provide an overview of the safety of the whole sample, as opposed to a highly selective and sensitive targeted analysis with mass spectrometry in which a small part of the composition of the sample is evaluated with no information of the toxicity of the compounds detected. Other teams have selected different combinations of techniques for assessing particle leakage. A recent study successfully detected silver nanoparticles

(10–100 nm) migrate from plastic food storage containers when in contact with food simulants using EDX [9]. This study also included other techniques [i.e., inductively coupled plasma-mass spectrometry (ICP-MS); single particle-ICP-MS (SP-ICP-MS), TEM-EDX] in determining the released nanoparticles, but could not quantitatively analyze particle shape, size, and agglomeration due to limited particle release [9].

Decision making within every stage of product development relies on the quality of analytical data obtained and the capacity to interpret the data. Simple and effective analytical methods are valuable in supporting the development of food related products and providing fast characterization for multidisciplinary teams of scientists. Although procedures involving fewer steps are largely preferred, attaining robust characterization of nanoparticles with a few steps is challenging given that they have complex behavior and exhibit a wide range of properties relevant to their use and toxicity that need to be determined. Moreover, the properties of nanoparticles may change over time due to their high reactivity. Several authors recognize that the presence of multiple influential factors that affect the properties of nanomaterials has created a bottle neck in gaining knowledge across different fields where nanotechnology is applied [10]. Nanoparticles are unusual analytes.

4.2 DETECTION OF NANOPARTICLES

In the development of products involving the use of nanomaterials, it is necessary to monitor nanoparticle features that are key to the intended function. This could be related to their structure or reactivity in water treatment applications (see Chapter 3) or their content of bioactive compounds in applications aiming to improve the nutritional value of foods (see Chapter 2). Crucially, it is vital to determine the size of the particulates as this is the characteristic feature in defining nanoparticles. According to the European Commission, the International Standards Organization (ISO) and American Society for Testing and Materials (ASTM), a nanoparticle by definition needs to have at least one of its dimensions measuring in the range of 1–100 nm [11–14]. However, there have been calls for the definition of a nanoparticle to broaden and include other relevant properties that are unique to nanoparticles. In this regard, the importance of internal and external domains in the nanorange, such as nanopores [15] and the inclusion of particles smaller than the currently adopted size range [11] are being evaluated. New experimental evidence and improved analytical capacity to characterize particles with dimensions below 1 nm will support decisions on widening the definition of nanoparticle. The upper limit may also be subject to review as new experimental evidence becomes available. The current guidelines also consider material as *nano* if 50% of the particles, in either free, aggregated, or agglomerated form, possess internal or external structures in the critical size range [11]. Consequently, the capacity to measure the size and size range of particles needs to be prioritized when deciding on the method and technique to characterize nanotechnology-based materials.

Besides particle size, composition, structure, surface chemistry, electronic properties (related to electric conductivity), and agglomeration are other important features that can define materials and affect their fate and toxicity in biological and environmental systems. One of the main challenges of sample treatment, required before the detection step, is the preservation of the particles in a form most closely resembling their original position within the matrix, that is, food products, without altering the properties that are characteristic to the particles as well as to their intended application. A good example of an attempt to preserve the properties of nanoparticles is the methodology used to extract metallic particles from plants [16]. This involved a mild and selective enzymatic digestion to release the nanoparticles from the plant tissue without chemical or physical (size and size distribution) alteration to the nanoparticles [16]. The extraction avoided the use of acidic digestion of the matrix to release nanoparticles; this could have affected the integrity of the silver nanoparticles.

One other characteristic of nanoparticles that affects the determination of particle size and size distribution is agglomeration. The tendency of nanoparticles to agglomerate is affected by properties such as particle composition, the concentration and type of surface coating, and the presence of cations in the media containing the particles. The latter can shield the superficial negative charges at the surface. Furthermore, the pH and ionic strength of the matrix within which the nanoparticles are contained can contribute to the formation of agglomerates. Unfortunately, the difficulty of maintaining the pH and ionic strength associated with the original sample during the analysis [17] presents the challenge of preserving particles in a form that represents their environment in the matrix, particularly with regard to agglomeration. The amount of loaded bioactive compound can also affect the size of the nanoemulsion. In a recent study, an increase in the amount of encapsulated substance was related to a decrease in the size of the nanoparticles. This also reduced the distance between the nanoparticle droplets and increased the viscosity of the emulsion [18].

The capabilities of the major detection techniques able to capture key characteristics of nanomaterials are summarized in Table 4.1. Among them, the best and almost only approach to assess specific surface area is through porosimetry (typically nitrogen porosimetry for micro- and mesoporous materials, or mercury porosimetry, for meso- and macroporous materials). It is also remarkable that microscopy techniques are so far the only ones that can differentiate between isolated particles and particle aggregates/agglomerates. Nonetheless, sample preparation steps and the isolation of particles from the matrix pose the challenge of maintaining the original particle dispersion found within the sample matrix. There is currently a range of techniques that are able to detect different sizes of particles but these techniques record agglomerates in the same way as primary individual particles without distinction. Likewise, the use of separation techniques such as field flow fractionation (FFF) or size exclusion chromatography, prior to detection, can separate by particle size but cannot distinguish agglomerates from primary particles. A cross-sectional view of the approaches currently used to characterize nanomaterials in various food related products found within recent papers (2015–2016) is shown in Table 4.2.

Table 4.1 Capability of detection techniques for the analysis of nanomaterials

Detection technique	Particle size	Composition	Internal structure	Morphology	Surface chemistry	Specific surface area	Electronic properties	Differentiation clusters/single particles	Quantitative analysis	High sensitivity
Microscopy (SEM, TEM, STEM, AFM, ASEM, ESEM, WetSEM)/ Microscopy-EDS	x	x	x	x				x		x
SLS, DLS	x							x	x	
Porosimetry			x			x			x	
UV-Vis, fluorescence	x			x			x	x	x	
XRD, XAS, XPS, EDX, EXAS, XANES	x	x	x	x	x		x		x	
Infrared, Raman		x			x		x			
ICP-MS or OES, AAS, GFAAS		x							x	x
SP-ICP-MS	x	x						x	x	x

AFM, atomic force microscopy; SEM, scanning electron microscopy; ASEM, atmospheric-SEM; ESEM, environmental-SEM; TEM, transmission electron microscopy; STEM, scanning transmission electron microscopy; EDS or EDX, energy dispersive spectroscopy X-ray spectroscopy; DLS, dynamic light scattering; SLS, static light scattering; UV-Vis spectroscopy, LIBD, laser-induced breakdown detection; AAS, atomic absorption spectroscopy; GFAAS, graphite furnace atomic absorption spectroscopy; XRD, X-ray diffraction; XAS, X-ray absorption spectroscopy; XRF, X-ray fluorescence; XPS, X-ray photoelectron spectroscopy; EXAFS, extended X-ray absorption fine structure spectroscopy; XANES, X-ray absorption near-edge spectroscopy, ICP-OES, inductively coupled plasma-optical emission spectroscopy; ICP-MS, inductively coupled plasma-mass spectrometry; SP-ICP, single particle-inductively coupled plasma mass spectrometry.

Table 4.2 Examples of analytical approaches used in recent studies to characterize nanoparticles in food related products

Nanoparticle	Matrix	Aim of the study	Sample treatment	Detection	Ref
Organic nanoparticles					
Lutein (73–96 nm)	SDS, sodium caseinate. Tween 20, 40, 60, 80	Optimization of the formulation of lutein nanodispersion	TEM: cover with fromvar film, air-dried, and stained with 2% phosphotungstic acid.	• TEM • DLS (backscattering at 173°) • Polydispersity index	[19]
Thymus daenensis (143 nm)	Tween 80, water, lecithin	Development of antibacterial nanoemulsion	Bacteria+ particles SEM imaging: fixation: 2.5% in PBS and de-hydrated in hydro-alcoholic solution. Coating with Au–Pd. Dilution before DLS. Centrifugation for stability test.	• SEM • DLS • UV-Vis: release of proteins • AAS release of K^+	[20]
Oregano oil (148 nm)	Oil, Tween 80, and water	Development of antibacterial nanoemulsion	Bacteria + particles SEM imaging: fixation: 2.5% glutaraldehyde in PBS, in 1% osmium tetroxide in PBS. De-hydratation in hydro-alcoholic solution. Coated with Au-Pd. Dilution before DLS.	• SEM of particles-bacteria • DLS • Z potential	[21]
D-limonene nanomulsion (31 nm)	D-limonene, sorbitane trioleate, polyoxyethylene (20) oleyl ether; ampicillin	Stability of nanoemulsions	TEM imaging: negatively stained with 1.5% (w/v) phosphotungstic acid. Excess liquid absorbed with filter paper. Dilution before DLS and turbidity measurements.	• TEM • DLS • Viscosity • Turbidity with UV-Vis	[22]

Carvacrol nanoemulsion (100 nm)	Carvacrol, miglyol, Tween 80, sodium citrate	Study antimicrobial properties		• DLS	[23]
Curcumin nanoemulsion (149 nm)	Curcumin, lactoferrin/alginate, corn oil, water	Stability of nanoemulsions in vitro	TEM imaging: coating with uranyl acetate and leave to dry. Dilution with solutions emulating stomach and intestine fluids before DLS and Z-potential measurement. Solutions for dilution: KCl-HCl, 10 mmol/L, pH 2 for stomach sample and phosphate buffer (10 mmol/L, pH 7 for initial and small intestine samples.	• TEM • DLS • Z potential • UV curcumin loading	[24]
Curcumin loaded nanoemulsion (18–30 nm)	Tween 80, phosphatidylcholine, ethyl oleate curcumin, bovine serum, ethanol, and water	Stability and interaction of curcumin nanoemulsions	Dilution prior to DLS TEM imaging: dilution before deposition onto mesh. Phosphotungstic acid was applied and its excess was removed by absorbing it with filter paper.	• DLS • Z potential • UV-Vis • Fluorescence • TEM • Refraction index • Viscosimetry • Surface tension	[25]
Medium chain triglycerides (10, 100 nm)	Medium chain triglycerides Tween 80, water, gelatin Gelatin (200)	Incorporation of nanoemulsions in hydrogels	No dilution prior DLS	• DLS • Temperature scanning analysis • Turbidity • Viscosity	[26]

(Continued)

Table 4.2 Examples of analytical approaches used in recent studies to characterize nanoparticles in food related products (*cont.*)

Nanoparticle	Matrix	Aim of the study	Sample treatment	Detection	Ref
Organic nanoparticles					
Peanut oil (73 nm)	Peanut oil, whey protein, β-lactoglobulin AB, SDS, X-triton-10	Optimization of the nanoemulsion		• DLS • Circular dichroism	[27]
Omega-3 (135 nm)	Fish and lemon oil, 100 mM acetic/acetate buffer pH 3, Tween 80, alginate	Influence of polysacharides on stability of omega-3 nanoemulsion	Confocal: dye with Nile red (hydrophobic) Static light scattering: dilution with acetic/acetate buffer pH 3 and stirred Z potential: sample diluted with 10 mM acetic/acetate buffer	• Confocal • Static light scattering • Z potential • Viscosity • Creaming index	[28]
D-limonene nanoemulsion (27 nm)	Sorbitane trioleate, polyoxyethylene oleyl ether, dropylene glycol, ampicillin, D-limonene, water	Stability of D-limonene nanoemulsion	TEM: negatively stained with 1.5% (w/v) phosphotungstic acid. Excess liquid absorbed with filter paper Dilution with water before DLS. Dilution with water prior turbidity measurement	• TEM • DLS • Turbidity: UV-Vis	[22]
Vitamine E nanoemulsion (110 nm)	Whey protein isolate Gum arabic Orange oil Vitamin E acetate	Formulation and stability of nanoemulsion		• Z potential • Static light scattering	[29]

Inorganic nanoparticles					
Licopene-nano-gold emulsion (17–25 nm) Gold (3–5 nm)		Inhibition of colon cancer cell growth by optimized nanoemulsion	Dilution before DLS TEM: Au: excess of sample absorbed with filter paper and dried overnight. Au-licopene: 2% phosphotungstic acid in pH 6.5-adjusted deionized water. The excess of stain was removed with filter paper. Cellular uptake: fixation of cells with 2.5% glutaraldehyde and 2% paraformaldehyde at 4°C. Postfixation with 1% osmium tetroxide, dehydrated with hydro-alcoholic solutions and acetone. Embed in Spurr's resin in an oven. Cut ultrathin sections and staining with 4% uranyl acetate and 0.2% lead citrate and mounting on a carbon-coated copper grid for TEM imaging	• TEM • Z potential • DLS • UV-Vis • HPLC-UV	[30]
Silver (60 nm)	In bacteria	Study the microbiological effect of nanosilver with different shapes	Pretreatment for TEM: a drop of *E.coli* containing Ag nanoparticles placed on a glow-discharged Formvar-coated copper grid for 2 min. The excess of liquid was absorbed with filter paper. The sample was treated with 1% phosphotungstic acid for 2 min.	• TEM • UV-Vis • XRD	[31]

(*Continued*)

Table 4.2 Examples of analytical approaches used in recent studies to characterize nanoparticles in food related products (*cont.*)

Nanoparticle	Matrix	Aim of the study	Sample treatment	Detection	Ref
Inorganic nanoparticles					
Silver (21 nm)	Plant tissue	Internalization of particles by plants	TEM: fixation with 2% paraformaldehyde and 2% glutaraldehyde in 0.05 M sodium cacodylate buffer for 4 h at 4°C. Dehydration with ethanol solutions and fixation in spur resin for 24 h at 72°C. Sections were cut with microtome and stained with 2% uranyl acetate and Reynolds lead citrate. SP-ICP-MS: clean plant tissues were blended with 2 mM citrate buffer (pH adjusted 3.5–10) and incubated with 5% solution of enzymes (cellulose, hemicellulase, and pectinase) for 24 h at 37°C. LOD 10 nm.	• TEM • SP-ICP-MS	[16]
Silver (>10 nm)	Food packaging	Migration of nanoparticles from food packaging into food	TEM: migration solution (1 mL) was centrifuged (30 min, 10.000 rpm) and evaporated until dry. Reconstituted sample (4 μL) was deposited onto the copper grid covered with carbon and dried. Total silver: plastic samples (0.25 g) were crushed < 2 mm. Microwave digestion with H_2SO_4 and HNO_3, 210°C, 20 min. ICP-MS for analysis of migrated silver particles: migrated solution incubated with HNO_3 (2:1) and diluted.	• TEM-EDS • ICP-MS • SP-ICP-MS	[9]

Silver (35 nm), copper (II) oxide (50 nm), and zinc oxide (20–30 nm)	Films for cheese	Retarding bacterial growth in cheese		• Quantification of particles with electro thermal atomic absorption spectrometry	[32]
Titanium dioxide (167 nm)	Flies	Toxicity study of titanium dioxide	ICP-MS. Digestion of the fly samples with aqua regia in microwave (2 h 120–195°C)	• TEM, AFM • ICP-MS • XPS • Isotherm: porosity and SBET • XRD	[33]
Titanium dioxide (21 nm)	Coating on stain steel	Physical stability and bactericidal properties of a titanium dioxide coating	—	• Variable pressure SEM • Scratch hardness • Adhesion strength	[34]
Aluminum nanoparticles (<50 nm and aggregates)		Preparation of a water filter	Leaching studies: trap potentially leached particles through nuclepore filters with a pore size of 50 nm. Freeze-drying filtrate samples and reconstituted with buffer for cell tests	• TEM • SEM • PIXE multielement analysis • MTS and LDH cell tests. • LC-MS	[5]

The type and average size of the particles and the analytical strategy used, including sample preparation and detection approaches, are given.

4.3 CHARACTERIZATION OF NANOMATERIALS USING NONDESTRUCTIVE TECHNIQUES

4.3.1 Microscopy

Microscopy techniques are among the various approaches that allow the characterization of nanomaterials with little alteration to the particles and their environment. SEM, TEM, confocal microscopy, atomic force microscopy (AFM), and their variants, are some of the few microscopic techniques that can provide direct and accurate measurement of particle size and in some cases detailed morphology (SEM and AFM). The size distribution and shape of particles obtained from micrographs can be further analyzed with processing software such as ImageJ [5,34–35] to achieve more comprehensive information.

SEM, TEM, and AFM are employed in cases where improved resolution is needed and the ability to distinguish agglomerates from primary nanoparticles. However, these techniques can only provide local information within the field of view and cannot provide detailed structural information imaging through dense specimens or cloudy liquids. In contrast, confocal microscopy is able to scan through a sample and provide a 3D image of some solids and liquids; however the technique has lower nanoparticle resolution compared to atomic and electron microscopy techniques [36]. In some cases confocal microscopy may not be able to resolve primary particles from agglomerates as most nanoparticles are smaller than the light wavelength used in this technique. A strength of confocal is that it allows for the imaging of samples in their natural state. This is advantageous for imaging nanoparticles in food and is important for understanding the fate of nanomaterials in cytotoxicity studies. An example of the importance of imaging particles in their real state in food can be found in a work investigating the effect of alginate on the stability of omega-3 nanodispersions. In this study, confocal microscopy enabled the examination of the structure of nanoemulsions at the microsize range (Fig. 4.1); this allowed for the determination of the conditions (% alginate) under which oil droplets were evenly distributed, stable against flocculation and when they started to coalesce in the presence of free alginate [28].

Sample preparation for SEM and TEM involves applying a sample droplet on a TEM grid or SEM sample holder where it is then left to dry. Following, it is coated with a fine layer of an electro-conductive metal (e.g., gold or palladium). This inhibits sample charging and improves the thermal stability of the sample, improving the signal obtained for topographic examinations. However, this type of sample preparation or dilution of the sample can affect interparticle interactions and may cause changes in particle size, shape, and agglomeration. Moreover, the drying process and vacuum conditions within the microscope may result in collapsed structures [25,37]. Additionally, there is also a risk of altering particles containing hydrophobic components, such as fat, when dehydrating, for instance with ethanol or acetone. The possible artifacts introduced when imaging

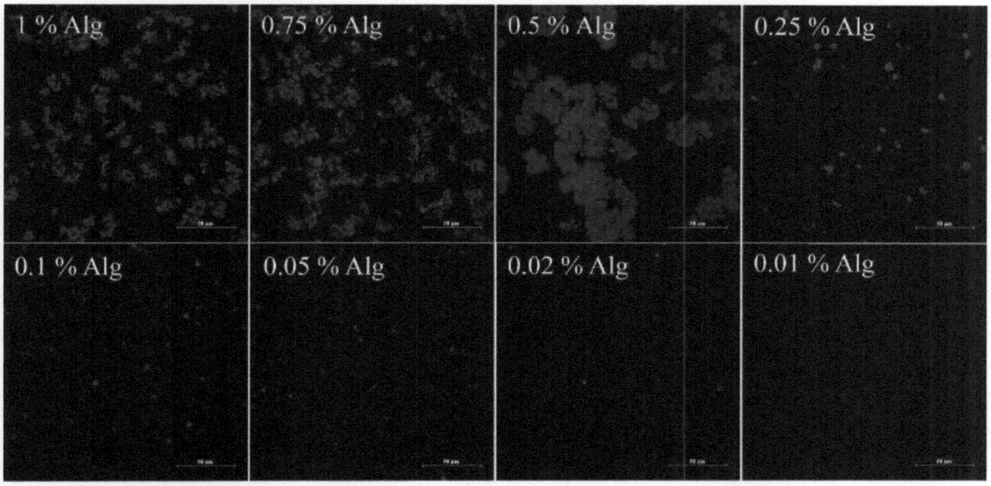

Figure 4.1 ***Confocal microscopy images of omega-3 nanoemulsions-sodium alginate mixtures after 16 days of storage time.*** *(Reproduced from L. Salvia-Trujillo, E.A. Decker, D.J. McClements, Influence of an anionic polysaccharide on the physical and oxidative stability of omega-3 nanoemulsions: antioxidant effects of alginate, Food Hydrocoll. 52 (2016) 690–698 [28] with permission from Elsevier).*

samples need to be taken into consideration as the morphology of the particles can be very important. In this regard, variations in toxicity have been observed from different nanoparticle shapes (i.e., spherocylindrical, pyriamidal, rod-like, spherical) [31,38]. Checks on the stability of the nanomaterial's size and structure, especially for organic particles by comparison with a reference sample, are recommended as these features are affected by different sample treatments and preparation processes. There are also TEM and SEM modalities that allow for the examination of nanoparticle under controlled pressure and humidity, representing a more realistic environment. Examples of these techniques include environmental-SEM (ESEM), atmospheric-SEM (ASEM), and Wet-SEM. The latter curtails the effect of the vacuum by imaging moist samples sealed in vacuum-resistant capsules [39]. These modalities can image hydrated samples and are therefore very appropriate for examining food samples and gain information about the presence of agglomerates [39]. Even though agglomerates may be above the nanorange, it is valuable to study them as they may exhibit vital properties due to their nanodomains or may de-agglomerate into their primary particles given that the attraction between particles may change with the external conditions. However, despite the high suitability of these microscopy techniques, their use is not yet widely adopted and traditional SEM and TEM approaches are mostly used (Table 4.2); SEM and TEM can resolve primary particles from agglomerates with the possibility of providing chemical composition such as in SEM/TEM-EDS. Confocal microscopes, unlike the newly developed ESEM, ASEM, and WetSEM technologies, are widely used in cell biology labs but despite this, its application for imaging nanoparticles in food products is not generalized, as it can be seen in Table 4.2.

In terms of resolution, AFM and TEM can provide subnanometer spatial resolution [40,41] and are very appropriate for imaging particles at subcellular level [16,31]. This can be important for carrying out characterizations that comply with the current European framework with regard to particle size. Although AFM can image a great variety of samples (e.g., food, tissues, ceramics, polymers), TEM requires ultrathin samples that can be penetrated by its electron beam [42]. In contrast to TEM, state-of-the art SEM offers less spatial resolution (of about 1 nm) [43] and only provides information from the morphology of the exposed surfaces of the particle. However, SEM can image bulk samples that can be cracked to allow the imaging of the inner surfaces of samples. Through imaging the cross-section of particles (i.e., crushed particles), information from the newly exposed internal morphology can be obtained (see abstract figure by Busquets et al. [44]). Cryo-SEM allows the snap freezing of a sample droplet that is then cracked to expose inner features. This process of cracking can be repeated to enable the analysis of different regions of the sample.

An especially relevant and challenging application for microscopy analysis is the determination of nanoparticle leachates from food contact materials. Depending on the level of leaching this problem can seem to be as difficult as looking for a needle in a haystack. However, TEM was found to be sufficient to detect migrated nanoparticles from all food packaging tested, using a minor preconcentration step, under EU regulation 10/2011 for plastic materials [45]. The silver particles migrating from packaging were mainly spherical and were found in three different states: dispersed, agglomerated, and still bound to the packaging, as depicted in Fig. 4.2. However, due to the low amount of nanoparticles released, quantitative measurements regarding size, shape, and agglomeration could not be carried out [9].

4.3.2 X-Ray spectroscopy

X-Ray techniques are used for the characterization of crystalline domains. In X-ray absorption spectroscopy (XAS), X-ray radiation interacts with electrons within the inner shells of atoms, which can result in the absorption of the initial radiation. However, XAS has the limitation of poor discrimination between neighbouring atoms in the periodic table. The interaction of X-rays with the ordered material can cause the expulsion of electrons from the inner core, which generates an ion with an excess of energy and a vacancy that was once occupied by the expelled electron. This vacancy can be filled by an electron of higher energy by emitting the excess of energy as an X-ray photon. This will present a longer wavelength than the starting radiation, which is the process involved in X-ray fluorescence (XRF).

On the other hand, the interaction of X-rays with electrons of a crystal can also result in dispersion of the radiation in a constructive or destructive manner, resulting

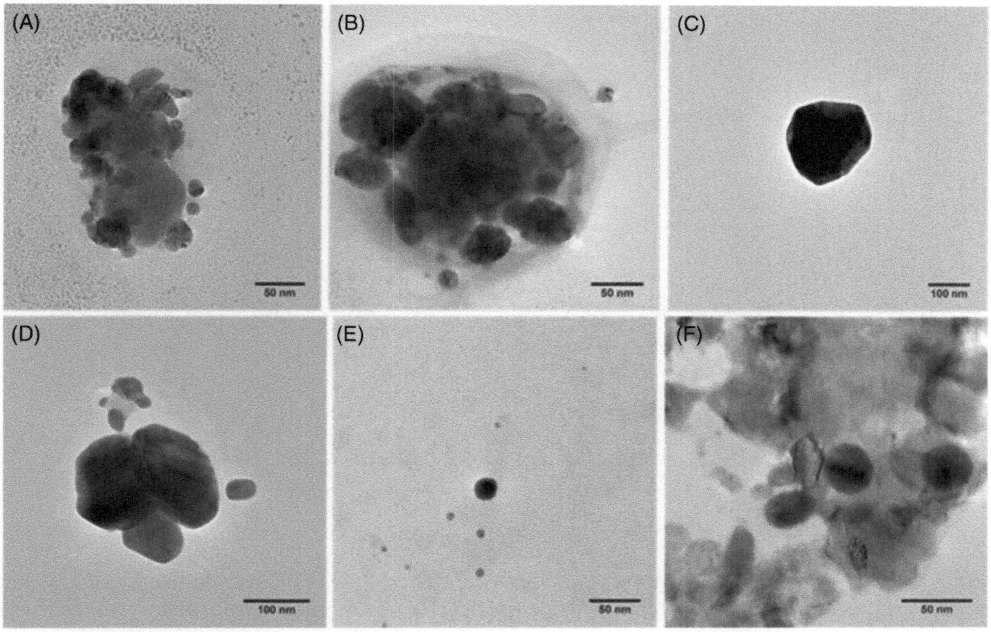

Figure 4.2 ***TEM results for migration solution samples after 10 days in 3% acetic acid.*** (A) Fresher Longer Miracle Food Storage bags. (B) Special nanosilver mother's milk pack. (C,D) The Original Always Fresh Containers. (E,F) Kinetic GoGreen Premium Food Storage Containers. All the presented images were analyzed by EDS and confirmed the presence of silver. *(Reproduced from A. Mackevica, M.E. Olsson, S.F. Hansen, Silver nanoparticle release from commercially available plastic food containers into food simulants, J. Nanopart. Res. 18 (2016) 1–11 [9] with permission from Springer).*

in diffraction (X-ray diffraction, XRD). The diffraction peaks can be converted into *d*-spacing, or distance between the ordered layers in a crystal, which are unique for each crystalline phase of a substance. This can be matched with a database to identify the atomic composition and structure of the material. The XRD pattern can also be used to differentiate between the different crystal facets of nanoparticles. For instance, Fig. 4.3 shows how silver nanowire, nanopsheres, and nanocubes, whose sizes were controlled by changing the amount of NaCl added during the synthesis of the nanoparticle, could be distinguished from each other using XRD [31].

X-Ray techniques provide information from the sample as a whole and are noninvasive. The data that they provide are related to the chemical speciation and structure of the inorganic material; the average size of the crystalline domain; and lattice deformation as a result of stress. It can be observed in Table 4.1 that X-ray techniques are the ones that can provide a broader range of characterization information. The Scherrer equation (4.1) relates the band width at half height in an XRD trace with the size of the ordered

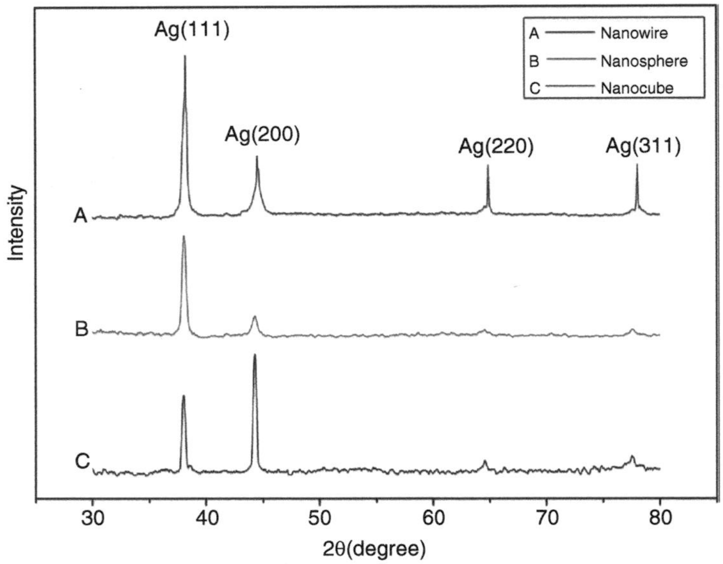

Figure 4.3 XRD patterns of silver nanowires (A), silver nanospheres (B), silver nanocubes (C). *(Reproduced from X. Hong, J. Wen, X. Xiong, Y. Hu, Shape effect on the antibacterial activity of silver nanoparticles synthesized via a microwave-assisted method, Environ. Sci. Pollut. Res. 23 (2016) 4489–4497 [31] with permission from Springer).*

domain, where τ is the size of the ordered domain; β is the band width at half height (measured in radians); κ is a constant related to the shape of the crystalline domain; λ is the wavelength used; and θ is the Bragg's angle or angle between the crystalline lattice and the incident and scattered radiation.

$$\tau = \frac{\kappa\lambda}{\beta\cos\theta} \tag{4.1}$$

However, there are some limitations that need to be considered when using this technique to measure particle size of a population of particles. The mean size of the ordered domains, τ, can be smaller than the size of particles because these may not be completely crystalline. X-Ray also has limited capacity in characterizing samples with important inhomogeneous strain or with particle sizes larger than a micrometer [46]. This technique can provide information related to the overall population of (crystalline) particle size, which is in line with the information required for nanomaterials according to the recommendations of the EU [11].

In a study where the same particles were characterized individually with TEM and XRD, both techniques gave similar size distributions for 10 nm [47] and 37 nm [48] particle size of copper and silver, respectively. Only TEM was able to characterize samples with particles of size of about 2 μm, the loss of validity of the analysis with XRD was attributed to the presence of smaller sized substructures [47]. The minimum sample amount required is another drawback in the use of XRD, which is in the order of tenths of grams, and therefore limits its use in cases involving low particle concentration as is commonly the case in particle leachates from packaging material. XRD is typically applied to characterize inorganic materials, and was found to be important in studying the mineral form of food grade titanium dioxide particles used in a toxicity study [33]. XRD can also be used in the characterization of organic structures. For instance, it was used to assess changes in the lipid crystallization process of nanostructured lipid carriers used to encapsulate lutein [49].

X-Ray can also be combined with SEM (i.e., SEM-EDS) to allow simultaneous analysis of the morphological features of particles and acquisition of the elemental composition of the exposed surface of the sample. However, this approach has sensitivity restrictions when compared with other techniques typically used for the analysis of metals such as ICP-MS, and its use in quantitative analysis requires samples with flat surface, which can be achieved by polishing the sample. Multielement analysis of nanoparticles, with higher sensitivity than EDS, in the ppm level, without the requirement for digestion of the particle, can be carried out with PIXE (particle-induced X-ray emission) [5,50].

A very powerful detection system that has also been coupled to microscopy is X-ray photoelectron spectroscopy (XPS). XPS provides information about the elemental composition, oxidation states, and chemical environment by measuring the kinetic energy of expelled electrons (photoelectrons) and relating it to the binding energy within the material. It was used to characterize the functional groups of food grade titanium dioxide nanoparticles before testing their toxicity with microorganisms. It was found through a specific band (532.5 eV) that the surface was partially covered with hydroxyl groups [33]. The combination of TEM-XPS provides a set-up that benefits from both the high resolution imaging and precise information of the size and shape of particles by TEM, with detailed data about the elemental composition of the surface of the particle, including oxidation states provided by XPS. As such, it can be used in studies such as the analysis of impurities on nanoparticles or defining the surface chemistry of nanoparticles. Table 4.2 gives an overview of the characterization methods used in recent works for both inorganic and organic nanomaterials relevant to food. Despite the superior properties of XPS, and XRD in terms of characterization data, as shown in Table 4.1, it is surprising that these techniques are not commonly employed in the characterization of nanoparticles related to food products.

4.3.3 Porosimetry

One of the distinguishing characteristics of nanomaterials is their high specific surface area in relation to their volume. This property is size dependent, especially for materials with low porosity. The EU recommends the measurement of specific surface area with nitrogen adsorption isotherms as an indicator of the presence of nanoparticles [11]. The specific surface area can be estimated using the Brunauer, Emmett and Teller (BET) adsoption model and porosimetry, which involves quantifying the number of gas molecules, typically nitrogen, that are needed to cover the surface of a material, which is in turn mathematically translated into surface area [51]. Typically, this measurement requires a sample amount providing a total specific surface area of at least 5 m^2. Lower specific surface area can still be measured using alternative adsorptive gases, such as krypton or xenon. This type of analysis is relevant to characterizing materials with nanodomains and relatively high surface area [33], or nanoporous materials, such as those included in food packaging or water filters [44].

The applications of porosimetry will be limited to the characterization of certain porous materials and nanoparticles which provide a total specific surface area in the sample cell above the sensitivity limit of the technique. Hence, while specific surface area can be measured for instance for 100 mg of carbon nanotubes before including them in a water filter, a sample of dry nanocapsules containing bioactive compounds with the same mass of 100 mg may not provide the minimum surface area that would guarantee accurate analysis. Although the amount of sample used can be increased to achieve the minimum specific surface area to make an accurate determination, the sample size can be limited to the capacity of the sample cell in the porosimeter. Likewise, if a food sample is potentially contaminated with particles that may have leached from ink or packaging material, the use of porosimetry will probably not provide enough sensitivity to detect the uptake of nitrogen by the particles leached to the food. Therefore it will not be possible to characterize food products that have been potentially contaminated with nanomaterials or most types of nanoingredients. Furthermore, given that the samples need to be completely dry before the adsorption of gas and the drying process may alter the structure (and possibly the porosity and surface area), this technique has limited application in the characterization of moist nanoingredients and food products. If microporosity (pores below 2 nm) becomes included in future definitions of nanoparticles, nitrogen porosimetry will become the most relevant approach of analysis as microscopy cannot provide information for the overall sample. In addition to specific surface area, porosimetry is unique in that it is able to measure the pore size distribution of the whole sample (Table 4.1). However, in practical terms, it is only applied infrequently at the stage of material development as highlighted in the compilation of studies in Table 4.2. Porosity data would be highly relevant when developing reactive or adsorbent materials for water purification, or packaging materials with barrier properties [44].

4.3.4 Static, dynamic light scattering, and laser Doppler microelectrophoresis

The size distribution of particles in suspension can be obtained with static and dynamic light scattering (DLS), whereas their stability can be determined with laser Doppler microelectrophoresis (Z-potential measurement). These three techniques are noninvasive and enable the analysis of particulates in liquid samples in their natural state.

Static light scattering (SLS) involves measuring the Brownian motion of particles, emulsions, and molecules in suspension by the illumination of the sample with a laser and measuring the intensity of the light scattered by the particles at a range of angles. The intensity of the scattered light is affected by particle size, molecular weight, shape of the particle, and the thermal energy of the system [52]. In DLS the fluctuations of light scattered by the particles in motion are measured at 90°. The fluctuations of the scattered light lead to the hydrodynamic diameter, as described by Stokes–Einstein equation [Eq. (4.2)], using the assumption that the particles are spherical [53].

$$D = \frac{kT}{c\pi\eta R} \tag{4.2}$$

In Eq. (4.2), D is the diffusion constant of the solute (particle) in a fluid; T is the absolute temperature of the system; R is the radius of the particles; η is the shear viscosity of the pure solvent; and c is a constant that depends on the selection of the hydrodynamic boundary conditions applied at the surface of the particles [54].

Consequently, the information obtained is that of a hydrodynamic sphere (solvated particle) and not the actual dimensional measurements of the particle size. As a result, particle size distributions determined with light scattering can differ from the particle sizes observed with microscopy. Nonetheless, a comparison between the particle size measurement obtained by DLS with the sample in suspension, and the direct imaging of the same dried sample by TEM, showed nonstatistical difference for nanoemulsions of curcumin with Tween 80/lecithin/ethyl oleate [25] and lycopene and nanogold [30].

DLS measurements can be carried out in batch mode or in a continuous following separation with size exclusion chromatography or filed flow fractionation [36]. At present, DLS analysis can be performed for particles as low as 0.3 nm in suspension [55], and can provide information comparable with TEM [41] in the lower working range of both techniques. When the acquisition of particle size distribution is carried out by dispersed-phase volume fraction (instead of by intensity), it is possible to observe if the smallest particles, which are typically the ones of most interest, represent a high volume fraction when there is not a monomodal distribution [27].

In SLS, DLS, and Z-potential measurements, particle concentrations need to be low to avoid multiple scattering effects that would lead to inaccurate results [52,53]. Diluting can be a solution but it can also introduce artifacts due to the modification of inter-particle interactions and the risk of affecting their structure. In some cases, particles can acquire markedly different shapes with different dilution rates [25].

SLS, DLS, and Z-potential measurements are not able to differentiate free primary particles from aggregates and agglomerates. As a result, high dispersion energy, that is, with ultrasonication, is first employed to disrupt particle clusters (agglomerates) and uniformly disperse the primary particles before analysis. It is however important to take note of and reduce the increase in sample heat resulting from prolonged sonication, as it can affect the thermal energy and Brownian motion of the particles. DLS is the technique most widely used to measure the size distribution of particles in suspensions; a summary of its applications from review of studies from 2015 and 2016 is given in Table 4.2. The popularity of the method is partly due to its cost effectiveness, quick and easy analysis, and the small size of DLS instruments.

4.3.5 UV-Vis, fluorescence, IR, and Raman in the analysis of nanoparticles

At the atomic scale, molecules with certain types of electrons can absorb UV-Vis radiation, and the portion of the transmitted (in UV-Vis, IR, and Raman) or emitted light (in fluorescence) can be used to identify and quantify the species that are interacting with a particular wavelength.

The interaction of light with particles is behind the early application of nanoparticles known in history, as exposed in Chapter 1, where it has been explained that nanoparticles of noble metals were included in stained glass for decorative and perhaps other practical functions. In the nanoscale range, when the frequency (or energy) of the incident radiation coincides with the collective oscillation of the conduction electrons of the nanoparticles, localized surface plasmon resonance can take place. In this phenomenon and at a particular wavelength, the extinction coefficient ε depends on the nanoparticle radius, shape, number of particles, composition, and the dielectric constant of the media, following the Mie theory [56]. Consequently, changes in the factors affecting ε can affect the wavelength of the absorbed or transmitted light. As a result, this enables localized surface plasmon resonance to evaluate particle radii, shape, and number of nanoparticles. However, this approach has drawbacks of low sensitivity and selectivity due to the broad spectral bands. An example of where this physical phenomenon has been used is in the investigation of the toxicity of silver particles with different shapes. Indeed, localized surface plasmon resonance was observed in the visible light for noble metal nanoparticles such as silver [57]. The presence of quasi-spherical silver nanoparticles could be determined because its spectra showed a predominant band at 430 nm; whereas the bands at 350 and 390 nm indicated the presence of nanowires; and the least energetic band (460 nm) indicated the presence of nanocubes, which was also confirmed with TEM imaging [31].

UV-Vis is one of the techniques that can provide information about agglomerates or aggregates (Table 4.1). In general, smaller particles account for a higher proportion of absorption than larger colloids that are inclined to scatter light. The scattering of light is very sensitive to particulate size, where clustering of particles increases the scattering. This causes the surface plasmon resonance to shift to lower energies (absorption bands shift to longer wavelengths) and the intensity of the extinction peak to decrease due to a reduction in the number of nanoparticles [58]. In addition, the presence of aggregates or agglomerates is typically characterized by the broadening of peaks with the second peak likely to appear at longer wavelengths when cluster formation is further increased. In a study testing the stability of curcumin nanoemulsions, UV-Vis helped to confirm that the particles of 10 and 100 nm could be kept stable for a month without affecting their Z-potential [25]. Another study assessing stability of particles used changes in UV-Vis absorption bands to monitor the aggregation of particles when these were prepared with similar capping agents [58]. Although the information that can be obtained with fluorescence and UV-Vis techniques is similar, like in molecular spectroscopy, fluorescence offers better sensitivity. However, as can be seen in Table 4.2, UV-Vis remains the technique of choice over fluorescence because the number of particle components that are fluorescent are more limited than those active in UV-Vis. A case where fluorescence has been recently used was in determining the interaction between curcumin and bovine serum albumin, which has fluorescent residues, in a nanoemulsion [25].

IR and Raman are complementary techniques that are generally used qualitatively to characterize organic particles or coatings onto inorganic particles. The coating can affect the stability of the particles with possible toxicity implications [59]. IR and Raman are very selective as different functional groups present bands at particular energies (or wavenumber) and are also further shifted by being in different molecular environments. Owing to this high selectivity, IR was employed for the quality control of the coating agent and reductant citric acid in silver nanoparticles [58] as well as terpenoids used as bio-reductant when plant extracts were used in the synthesis of the same metal particles [48]. A particularly interesting work used IR to optimize the composition of a film from which antimicrobial agents had migrated to meat as well as to study the changes to the film during its use [4]. Although IR is frequently used to characterize nanoparticles after their synthesis, it is not commonly used to analyze nanoparticles within their primary matrix as nanoparticles need to be isolated from matrix to be characterized. Nevertheless, others have achieved using IR to study molecules within a complex matrix with one good example being the use of IR assisted determination of the location of the curcumin within the micelles in a nanoemulsion [25]. Although spectroscopic techniques, UV-Vis, fluorescence, Raman and IR, provide simple and quick monitoring of the stability of nanoparticle dispersions and composition, the selectivity and sensitivity of these techniques are limited when the analytes are within a complex matrix such as food and steps toward the isolation of the nanoparticles are needed.

4.4 CHARACTERIZATION OF NANOMATERIALS USING DESTRUCTIVE TECHNIQUES

The inclusion of nanomaterials in the production of food containers and utensils affects properties of both the material and the nanoingredient in terms of its stability and toxicity. For instance, nanomaterials such as zinc oxide and silver can dissolve and release ions in solution that travel to the food products. Although this phenomenon is exploited and tailored toward imparting antimicrobial properties into the material [32,60,61], other nanomaterials or higher concentration of otherwise safer nanomaterials may be toxic and/or compromise the integrity of the food. As such, quantification of the migrated ions and nanomaterials to the food is essential although this is not always possible due to limitations in sensitivity of the techniques capable to carry out such determination [9]. Besides sensitivity issues, the instruments used often required a particular minimum amount of sample to enable the collection of robust quantitative data. As such, it is not possible to obtain quantitative results from either smaller sample sizes that the requirement of the techniques or samples containing too limited amounts of particles, situation that can occur when the food becomes contaminated as a result of some minor loss of nanomaterials from the food storage container, packaging material, or filter unit (in the case of liquids).

In contrast, microscopy is one of the techniques that require only a few microliters of sample for analysis, with TEM and SEM being capable of analyzing even a single nanoparticle within the field of view (Table 4.2). Despite this clear strength, localized portions of the sample are imaged, which is again a thread to the representativeness of the information that the technique provides. Inductively coupled plasma (ICP) is conventionally used in the analysis of metals at atomic level. It makes possible to provide multi-elemental analysis of most elements in the periodic table with high sensitivity detection (in the range of low parts per billion) and is commonly used in environmental analysis. This technique offers superior capabilities for the analysis of metals nanomaterials as well. The amount of sample required in the analysis with this technique is higher than in DLS or microscopy: several milliliters are needed, which makes the results obtained more representative, and also improves the relatively high limit of detection associated with other techniques such as X-ray (Table 4.1). ICP-MS has been used for quantifying the total amount of silver nanoparticle in food containers (<0.2–$1.4\ \mu g/cm^2$) after the acidic digestion of representative samples [9]. It has also been used to quantify food grade titanium dioxide particles ingested by a few specimens only (3–4 flies which ingested $1002 \pm 105\ \mu g$ titanium dioxide/g fly body weight after exposure to 2 mg titanium dioxide/L in feeding medium for 4 days) and digestion of the specimens immersed in aqua regia with pressurized microwave extraction [33].

In ICP-MS, typical acquisition dwell times are in the region of milliseconds. Although this is suitable for the analysis of particles once they have been dissolved, it has been reported to be too slow for detecting individual nanoparticle events in the plasma, which have been estimated to have a duration of 0.5 ms [62]. This drawback can be overcome by using single particle-inductively coupled plasma mass spectrometry (SP-ICP-MS) which has lower dwell times than ICP-MS, which is its predecessor technique. This enables the technique to differentiate between the signal from particles, based on pulse signals from the particles entering the plasma (transient peaks or metals), and the signal of the dissolved metal, which appears in the background [63]. SP-ICP-MS has gained popularity among researchers owing to its ability to determine the number of particles or aggregates from the frequency of the signals, and determine particle diameter and mass of the elements making up the particle from the intensity of the signal (Table 4.2). The reliability of SP-ICP-MS was demonstrated by a study that analyzed silver nanoparticles (20 nm) extracted from plant tissues and produced results that were consistent with the results obtained from TEM [16]. However, the minimum detectable particle size of silver nanoparticles that could be detected with SPS-ICP-MS tissue was 10 nm [16] and the inability to quantify particles in the smaller size range (approximately 1–20 nm) currently used in food technology has been pointed out as limitation of this technique. Nonetheless, a recent study was able to quantify silver nanoparticles (24 nm) leached from food containers to food stimulants at concentrations of $>1.9 \times 10^6$ particles/L, with silver nanoparticles detection in 11 out of 12 samples. This demonstrates that SPS-ICP-MS is also an appropriate technique in the study of the migration of nanomaterials from materials and products [9]. Indeed, the high sensitivity of this technique makes it the most appropriate for monitoring the fate of nanoparticles in cells of living things [64].

4.5 OVERVIEW OF RECENT METHODS FOR THE ANALYSIS OF ENGINEERED NANOMATERIALS AND FUTURE TRENDS

Engineered nanoparticles can be in the form of inorganic particles (i.e., metals and metal oxides nanoparticles) or organic particles such as those based on natural and synthetic polymers produced by emulsions and or crosslinking by ionic or covalent interaction. Both these forms can further exist as free primary particles or as aggregates and agglomerates. In addition, they can be in the form of encapsulated or bound (in composites) inorganic particles [65]. The aim of the studies including the nanoingredients or nanomaterials in different type of formulations and composites reviewed in Table 4.2 has been to investigate improvements in technical properties of the product or to study the fate of nanomaterials included in the food related products. Studies related to antimicrobial properties are among the most common themes of research in food nanotechnology: antimicrobial properties have been investigated in a number of nanoingredients such as silver [31], titanium dioxide [34], and natural products such as the terpenoid carvacrol [23] or oils [21]. Samples with particle size slightly above 100 nm have been included in Table 4.2 to keep an open view as the size of the particles found in food items can vary following changes in the surrounding area (i.e., food matrix) which can happen during the measurement [17].

ESEM, WetSEM, and ASEM are microscopy modalities most recommended because of the high extent of preservation of the particle environment in the sample being imaged and the direct analysis offered. Nevertheless, these are not widely used yet (Table 4.2), possibly because of their cost and technical skills needed for this measurements as compared to conventional measurements with SEM, but it is expected that they will be progressively introduced in the development of food products containing nanotechnology.

On the contrary, measurement of particle size with DLS has been the most extensively used approach. In DLS, SPS, and Z-potential, authors face the dilemma of diluting or not the samples before the measurement, and both options are followed, as it can be seen in Table 4.2. Not diluting the samples may lead to multiple scattering effects, but diluting the sample may change interparticle interaction and agglomeration. When using DLS, SPS, or carrying out measurements of Z-potential, it can be recommended to sonicate the sample immediately before the measurement, used volume-weighed distributions and, if necessary, dilute samples with extracting solution as similar as possible to the food matrix (pH, ionic strength, chemical composition), or cell culture, where relevant. Standardized methodology for the analysis of nanomaterials in food products reflecting the original conditions, without causing alterations in the particles, is not available. Actions toward harmonizing methodology are needed, and for that reference materials need to be prepared and made available, and laboratories need to participate in intercomparison exercises. The absence of harmonized methodology has led to the current diversity of sample treatments carried out before the particle size measurements.

The coating of particles increases their colloidal stability in suspension by increasing electrostatic or steric repulsion. Increasing electrostatic repulsion is often carried out with citrate in synthetic processes, whereas enhancing steric repulsion is achieved, in both organic and inorganic particles, with bigger molecules such as polymers (i.e., Tween [19,21], or both, by using polyelectrolytes (i.e., casein [19], gum Arabic [29], whey protein [29], lactoferrin, alginate [24]). Particles that present steric stabilization will preserve more easily their agglomeration state during sample treatment because they are less affected by changes caused by ionic strength than uncoated particles or particles with coating for improving electrostatic stabilization only [58]. A natural coating can also be gained by adsorption of dissolved organic matter onto the particle, which may improve their stability [66]. Currently, the characteristics and stability of this coating in analytical procedures is a property that is being overlooked and rarely addressed. As it can be seen in Table 4.2, IR or Raman has been rarely used and solid state NMR or mass spectrometry analysis of the coating has not been attempted.

Similarly, separation techniques which are an effective approach to separate nanoparticles for their determination such as size exclusion chromatography, FFF, capillary electrophoresis, or hydrodynamic chromatography, after washing steps to remove possible macromolecules, are not widely applied despite the improvements in quantitative analysis that they can offer. These techniques have been recently discussed and reviewed for the analysis of nanoparticles [36]. The underlying cause of this could be that both the expertise needed for setting up the separation of nanoparticles from food products and time required to carry out such determination cannot compete with other easy-to-use approaches such as DLS. Another cause of this situation can be that the instrumentation to carry out the separation of nanoparticles is not readily available in material, food or life sciences labs, which are all important contributors in the development of emerging nanotechnologies in food science.

The analysis of nanomaterials in food products is of utmost importance due to the major advances that they can provide and the urgent need to know more about the properties and behavior in food related products, living organisms, and the environment. The changing nature of such analytes, due to their high reactivity and interaction with the environment surrounding the particles, makes their analysis a challenge compared to the determination of atoms or molecules, and food matrices are well known to be challenging themselves. A range of techniques available can capture relevant characteristics of nanomaterials but no technique can provide all features that are needed. Furthermore, analytical strategies need to progress toward quantitative determinations and the development of harmonized and validated methodology is needed. Presently, the application of a combination of analytical approaches is the recommended way to move forward in the analysis of nanomaterials in food products.

ACKNOWLEDGMENTS

The FP7 Marie Curie program (project nos 274985 and 230676) is acknowledged for having fostered interdisciplinary skills relevant to this chapter. Dr Lubinda Mbundi (Blond McIndoe Research foundation) and Patrick Melia (Kingston University) are acknowledged for critical discussions.

REFERENCES

[1] DTU Environment, D.C. Council, The Nanodatabase. Available from: http://nanodb.dk/en/search-database/
[2] A.A. Keller, S. McFerran, A. Lazareva, S. Sangwon, Global life cycle releases of engineered nanomaterials, J. Nanopart. Res. 15 (2013) 1–17.
[3] R. Damoiseaux, S. George, M. Li, S. Pokhrel, Z. Ji, B. France, T. Xia, E. Suarez, R. Rallo, L. Mädler, Y. Cohen, E. Hoek, A. Nel, No time to lose—high throughput screening to assess nanomaterial safety, Nanoscale 3 (2011) 1345–1360.
[4] M.V. Dias, N. De Fátima, F. Soares, S.V. Borges, M.M. De Sousa, C.A. Nunes, I.R.N. De Oliveira, E.A. a Medeiros, Use of allyl isothiocyanate and carbon nanotubes in an antimicrobial film to package shredded, cooked chicken meat, Food Chem. 141 (2013) 3160–3166.
[5] L. Önnby, C. Svensson, L. Mbundi, R. Busquets, A. Cundy, H. Kirsebom, Γ-Al_2O_3-based nanocomposite adsorbents for arsenic(V) removal: assessing performance, toxicity and particle leakage, Sci. Total Environ. 473-474 (2014) 207–214.
[6] X. Zhang, C.R. Li, W.C. Wang, J. Xue, Y.L. Huang, X.X. Yang, B. Tan, X.P. Zhou, C. Shao, S.J. Ding, J.F. Qiu, A novel electrochemical immunosensor for highly sensitive detection of aflatoxin B1 in corn using single-walled carbon nanotubes/chitosan, Food Chem. 192 (1992) (2016) 197–202.
[7] Promega, CellTiter 96® AQueous non-radioactive cell proliferation assay, in: Promega (Ed.), Technical Bulletin No. 169, Promega Cooperation, Madison, USA, 2001.
[8] Promega, CytoTox 96® non-radioactive cytotoxicity assay. Instructions for use of product G1780., Madison, USA, 2012.
[9] A. Mackevica, M.E. Olsson, S.F. Hansen, Silver nanoparticle release from commercially available plastic food containers into food simulants, J. Nanopart. Res. 18 (2016) 1–11.
[10] E.C. Díaz, M.V. Cases, Analytical methodologies for nanotoxicity assessment, TrAC, Trends Anal. Chem. 84 (part A) (2016) Available from: http://dx.doi.org/10.1016/j.trac.2016.03.007.
[11] The European Commision, Commission recommendation of 18 October 2011 on the definition of nanomaterial, Off. J. Eur. Union (2011) 2011/696/EU, L275/38-L275/40.
[12] E. Commission, Definition of a nanomaterial, 2015. Available from: http://ec.europa.eu/environment/chemicals/nanotech/faq/definition_en.htm

[13] ISO, International Organization of Standardization, Nanotechnologies—Vocabulary—Part 2 Nano-objects, Geneva, Switzerland, 2015.

[14] A. International, ASTM E2456-06 (2012), Standard Terminology Relating to Nanotechnology, West Conshohocken, PA, USA, 2012. www.astm.org.

[15] ISO, Nanotechnologies—Vocabularies for science, technology and innovation indicators, ISO/TS 18110:2015(en), 2015. Available from: https://www.iso.org/obp/ui/#iso:std:iso:ts:18110:ed-1:v1:en:term:2.8

[16] D. Bao, Z.G. Oh, Z. Chen, Characterization of silver nanoparticles internalized by arabidopsis plants using single particle ICP-MS analysis, Front. Plant Sci. 7 (2016) 1–8.

[17] A.M. EL Badawy, T.P. Luxton, R.G. Silva, K.G. Scheckel, M.T. Suidan, T.M. Tolaymat, Impact of environmental conditions (pH, ionic strength, and electrolyte type) on the surface charge and aggregation of silver nanoparticles suspensions, Environ Sci. Technol. 44 (4) (2010) 1260–1266.

[18] G.D. Chanana, B.B. Sheth, Particle size reduction of emulsions by formulation design-II: effect of oil and surfactant concentration, J. Pharm. Sci. Technol. 49 (1995) 71–76.

[19] T.B. Tan, N.S. Yussof, F. Abas, H. Mirhosseini, I.A. Nehdi, C.P. Tan, Forming a lutein nanodispersion via solvent displacement method: the effects of processing parameters and emulsifiers with different stabilizing mechanisms, Food Chem. 194 (2016) 416–423.

[20] R. Moghimi, L. Ghaderi, H. Rafati, A. Aliahmadi, D.J. McClements, Superior antibacterial activity of nanoemulsion of Thymus daenensis essential oil against *E. coli*, Food Chem. 194 (2016) 410–415.

[21] K. Bhargava, D.S. Conti, S.R.P. da Rocha, Y. Zhang, Application of an oregano oil nanoemulsion to the control of foodborne bacteria on fresh lettuce, Food Microbiol. 47 (2015) 69–73.

[22] P.H. Li, W.C. Lu, Effects of storage conditions on the physical stability of D-limonene nanoemulsion, Food Hydrocoll. 53 (2015) 1–7.

[23] K.S. Landry, S. Micheli, D.J. McClements, L. McLandsborough, Effectiveness of a spontaneous carvacrol nanoemulsion against *Salmonella enterica Enteritidis* and *Escherichia coli* O157:H7 on contaminated broccoli and radish seeds, Food Microbiol. 51 (2015) 10–17.

[24] A.C. Pinheiro, M.A. Coimbra, A.A. Vicente, In vitro behaviour of curcumin nanoemulsions stabilized by biopolymer emulsifiers—effect of interfacial composition, Food Hydrocoll. 52 (2016) 460–467.

[25] K. Kaur, R. Kumar, S.K. Mehta, Nanoemulsion: a new medium to study the interactions and stability of curcumin with bovine serum albumin, J. Mol. Liq. 209 (2015) 62–70.

[26] J. Komaiko, D.J. McClements, Food-grade nanoemulsion filled hydrogels formed by spontaneous emulsification and gelation: Optical properties, rheology, and stability, Food Hydrocoll. 46 (2015) 67–75.

[27] X. Zhang, Z.Z. Haque, Generation and stabilization of whey-based monodisperse nanoemulsions using ultra-high-pressure homogenization and small amphipathic co-emulsifier combinations, J. Agric. Food Chem. 63 (2015) 10070–10077.

[28] L. Salvia-Trujillo, E.A. Decker, D.J. McClements, Influence of an anionic polysaccharide on the physical and oxidative stability of omega-3 nanoemulsions: antioxidant effects of alginate, Food Hydrocoll. 52 (2016) 690–698.

[29] B. Ozturk, S. Argin, M. Ozilgen, D.J. McClements, Formation and stabilization of nanoemulsion-based vitamin E delivery systems using natural biopolymers: Whey protein isolate and gum arabic, Food Chem. 188 (2015) 256–263.

[30] R.F.S. Huang, Y.J. Wei, B.S. Inbaraj, B.H. Chen, Inhibition of colon cancer cell growth by nanoemulsion carrying gold nanoparticles and lycopene, Int. J. Nanomed. 10 (2015) 2823–2846.

[31] X. Hong, J. Wen, X. Xiong, Y. Hu, Shape effect on the antibacterial activity of silver nanoparticles synthesized via a microwave-assisted method, Environ. Sci. Pollut. Res. 23 (2016) 4489–4497.

[32] F. Beigmohammadi, S.H. Peighambardoust, J. Hesari, S. Azadmard-Damirchi, S.J. Peighambardoust, N.K. Khosrowshahi, Antibacterial properties of LDPE nanocomposite films in packaging of UF cheese, LWT-Food Sci. Technol. 65 (2016) 106–111.

[33] B. Jovanovic, V.J. Cvetkovic, T.L. Mitrovic, Effects of human food grade titanium dioxide nanoparticle dietary exposure on Drosophila melanogaster survival, fecundity, pupation and expression of antioxidant genes, Chemosphere 144 (2016) 43–49.

[34] V.K. Yemmireddy, Y.C. Hung, Effect of binder on the physical stability and bactericidal property of titanium dioxide (TiO2) nanocoatings on food contact surfaces, Food Control 57 (2015) 82–88.

[35] R.D. Glover, J.M. Miller, J.E. Hutchison, Generation of metal nanoparticles from silver and copper objects: Nanoparticle dynamics on surfaces and potential sources of nanoparticles in the environment, ACS Nano 5 (2011) 8950–8957.
[36] L. Mbundi, H. Gallar-Ayala, M.R. Khan, J.L. Barber, S. Losada, R. Busquets, Advances in the analysis of challenging food contaminants: Nanoparticles, bisphenols, mycotoxins, and brominated flame retardants, Adv. Mol. Toxicol. 8 (2014) 35–105.
[37] R.L.D. Whitby, V.M. Gun'ko, A. Korobeinyk, R. Busquets, A.B. Cundy, K. László, J. Skubiszewska-Zięba, R. Leboda, E. Tombácz, I.Y. Toth, K. Kovacs, S.V. Mikhalovsky, Driving forces of conformational changes in single-layer graphene oxide, ACS Nano 6 (2012) 3967–3973.
[38] C.M. Beddoes, C.P. Case, W.H. Briscoe, Understanding nanoparticle cellular entry: a physicochemical perspective, Adv. Colloid Interface Sci. 218 (2015) 48–68.
[39] P. Luo, I. Morrison, A. Dudkiewicz, K. Tiede, E. Boyes, P. O'Toole, S. Park, A.B. Boxall, Visualization and characterization of engineered nanoparticles in complex environmental and food matrices using atmospheric scanning electron microscopy, J. Microsc. 250 (1) (2013) 32–41.
[40] E. Balnois, G. Papastavrou, K.J. Wilkinson, Force microscopy and force measurements of environmental colloids, in: K.J. Wilkinson, J.R. Lead (Eds.), Environmental Colloids and Particles: Behaviour, Separation and Characterisation, vol. 10, John Wiley & Sons Ltd, Chichester, UK, 2007, pp. 405–467 Chapter 9.
[41] JEOL, JEM-ARM300F Grand ARM Transmission Electron Microscope, JEOL Solut. Innov. 2016. Available from: http://www.jeolusa.com/PRODUCTS/Transmission-Electron-Microscopes-TEM/300-kV/JEM-ARM300F
[42] Z. Yu, B. Xiao, W. Huang, P. Peng, Sorption of steroid estrogens to soils and sediments, Environ. Toxicol. Chem. 23 (2004) 531–539.
[43] JEOL, JSM-7800F PRIME Extreme-resolution Analytical Field Emission SEM. 2016. Available from: http://www.jeolusa.com/PRODUCTS/Scanning-Electron-Microscopes-SEM/FE-SEM/JSM-7800F-PRIME
[44] R. Busquets, O.P. Kozynchenko, R.L.D. Whitby, S.R. Tennison, A.B. Cundy, Phenolic carbon tailored for the removal of polar organic contaminants from water: a solution to the metaldehyde problem?, Wat. Res. 61 (2014) 46–56.
[45] Commission European, Commission Regulation (EU) No. 10/2011 of 14 January 2011 on plastic materials and articles intended to come into contact with food, Off. J. Eur. Communities 50 (2011) 1–89.
[46] Y. Zhao, J. Zhang, Microstrain and grain-size analysis from diffraction peak width and graphical derivation of high-pressure thermomechanics, J. Appl. Crystallogr. 41 (2008) 1095–1108.
[47] Y. Zhong, D. Ping, X. Song, F. Yin, Determination of grain size by XRD profile analysis and TEM counting in nano-structured Cu, J. Alloys Compd. 476 (2009) 113–117.
[48] T.C. Prathna, N. Chandrasekaran, A.M. Raichur, A. Mukherjee, Kinetic evolution studies of silver nanoparticles in a bio-based green synthesis process, Colloids Surf., A 377 (2011) 212–216.
[49] C.H. Liu, C.T. Wu, Optimization of nanostructured lipid carriers for lutein delivery, Colloids Surf., A 353 (2010) 149–156.
[50] T. Johansson, R. Akselsson, S. Johansson, X-ray analysis: elemental trace analysis at the 10–12 g level, Nucl. Instrum. Methods 84 (1970) 141–143.
[51] S. Gregg, K.S.W. Sing, Adsorption, Surface Area, and Porosity, second ed., Academic Press, London, (1995).
[52] LS instruments, Static Light Scattering. Available from: http://www.lsinstruments.ch/technology/static_light_scattering_sls/
[53] Malvern Instruments Ltd, A basic guide to particle characterization, Inform White Paper (2012) 1–26.
[54] J.R. Schmidt, J.L. Skinner, Brownian motion of a rough sphere and the Stokes–Einstein law, J. Phys. Chem. B 108 (2004) 6767–6771.
[55] Marlvern Instruments Limited, Accuracy and reproducibility in automated dynamic light scattering measurements, 2015.
[56] M. Sonntag, J. Klingsporn, A. Zrimsek, B. Sharma, L. Ruvuna, R. Van Duyne, Molecular plasmonics for nanoscale spectroscopy, Chem. Soc. Rev. 43 (2014) 1230–1247.
[57] M. Kerker, The optics of colloidal silver: something old and something new, J. Colloid Interface 105 (1985) 297–314.

[58] T.C. Prathna, N. Chandrasekaran, A. Mukherjee, Studies on aggregation behaviour of silver nanoparticles in aqueous matrices: Effect of surface functionalization and matrix composition, Colloids Surf., A 390 (2011) 216–224.
[59] C. Marambio-Jones, E. Hoek, A review of the antibacterial effects of silver nanomaterials and potential implications for human health and the environment, J. Nanopart. Res. 12 (2010) 1531–1551.
[60] C. Levard, E.M. Hotze, G.V. Lowry, G.E. Brown, Environmental transformations of silver nanoparticles: impact on stability and toxicity, Environ. Sci. Technol. 46 (2012) 6900–6914.
[61] N. Milani, M.J. McLaughlin, S.P. Stacey, J.K. Kirby, G.M. Hettiarachchi, D.G. Beak, G. Cornelis, Dissolution kinetics of macronutrient fertilizers coated with manufactured zinc oxide nanoparticles, J. Agric. Food Chem. 60 (2012) 3991–3998.
[62] J.W. Olesik, P.J. Gray, Considerations for measurement of individual nanoparticles or microparticles by ICP-MS: determination of the number of particles and the analyte mass in each particle, J. Anal. At. Spectrom. 27 (2012) 1143–1155.
[63] A. Hineman, C. Stephan, Effect of dwell time on singleparticle inductively coupled plasma mass spectrometry data acquisition quality, J. Anal. At. Spectrom. 29 (2014) 1252–1257.
[64] Y. Arai, T. Miyayama, S. Hirano, Difference in the toxicity mechanism between ion and nanoparticle forms of silver in the mouse lung and in macrophages, Toxicology 328 (2015) 84–92.
[65] R.Y. Yada, N. Buck, R. Canady, C. DeMerlis, T. Duncan, G. Janer, L. Juneja, M. Lin, D.J. Mcclements, G. Noonan, J. Oxley, C. Sabliov, L. Tsytsikova, S. Vázquez-Campos, J. Yourick, Q. Zhong, S. Thurmond, Engineered nanoscale food ingredients: evaluation of current knowledge on material characteristics relevant to uptake from the gastrointestinal tract, Compr. Rev. Food Sci. Food Saf. 13 (2014) 730–744.
[66] S.M. Rodrigues, T. Trindade, A.C. Duarte, E. Pereira, G.F. Koopmans, P.F.A.M. Römkens, A framework to measure the availability of engineered nanoparticles in soils: trends in soil tests and analytical tools, TrAC—Trends Anal. Chem. 75 (2016) 129–140.

CHAPTER FIVE

Bioavailability of Nanomaterials and Interaction With Cells

Joseph Lacey
Rayner Intraocular Lenses Limited, Worthing, United Kingdom

Contents

5.1 INTRODUCTION

The application of nanoscale particulates in food products has revolutionized the industry through greatly improving the control of food quality and food packaging, topics that have been addressed in depth in Chapters 2, 7, and 8 respectively. Nanomaterials are also widely used in other goods such as in personal care products [1]. The average western person's gastrointestinal system (GIS) is exposed to billions of nanoparticles every day through food and cosmetic products, many of which may enter the body [2]. However, little is commonly known about the toxicity of these materials and how they interact with the biological systems with which they encounter. Central to understanding the potential risks of toxicity is understanding the bioavailability of these materials and how they can directly enter the body and disrupt or interfere with the function of its different cellular mechanisms.

The bioavailability of nanomaterials is highly dependent on particulate size, porosity, chemical composition, density, and the biological system with which it is interacting [3,4]. Therefore to understand some of the complex processes that can determine a nanomaterials' bioavailability it is simplest to follow the journey they take through the human body, and how they may interact with the tissues that they encounter. To understand the bioavailability of a given nanomaterial with a given tissue we must first explore the mechanisms by which cells and tissues can interact with and transport nanoparticles.

5.1.1 Mechanisms of cellular and extracellular transport of nanoparticles

There are two main classes of transport that affect nanoparticles; these are transcellular and paracellular [5]. As the name suggests, transcellular transport is a process by which the cells of a tissue utilize a mechanism of transport through the cell [5,6] whereas a paracellular transport is a process that occurs due to free passage or physical effect that passes not through cells but through extracellular spaces [7].

Among transcellular processes, there are a number of different types of transport, most notably endocytosis; active transport; and facilitated diffusion (Fig. 5.1) [6]. Specifically, there are three different main forms of endocytosis: phagocytosis, pinocytosis, and receptor-mediated endocytosis.

Phagocytosis is a process of transport that functions by a cell engulfing and encapsulating the nanoparticle within the cell [8]. This type of process is most commonly seen in cells of immune function such as macrophages as a mechanism of destroying a foreign body, virus, or microorganism infecting the host [8]. The production of macrophage is highly regulated but increases in response to immune activation and inflammation of tissues [9].

Pinocytosis, however, is common to almost all eukaryotic cell types and is often described as "cell drinking" where cells form invaginations of the cell membrane to create a vesical filled with extracellular fluid [10]. This process allows the cell to take up many nutrients from the extracellular fluid and also some small proteins and particulates [11]. Due to its nonspecific nature, the process of pinocytosis is thought to be one of the primary mechanisms of nanoparticle transport, with numerous publications demonstrating the ready transport of different nanomaterials using this process [10]. These include gold

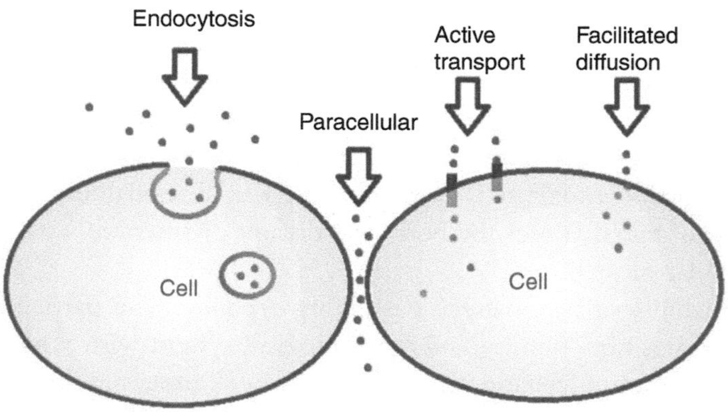

Figure 5.1 ***Mechanisms of cellular transport.***

[12,13], quantum dots [14], and hydroxyapatite nanoparticles [15]. Although this process is thought to be the main mechanism of nanoparticle transport, other mechanisms are still considered to be highly probable due to the extremely small sizes being able to bypass tightly controlled active transport processes.

Active transport is a process by which membrane proteins can selectively transport molecules into and out of a cell using energy-dependent mechanisms [16]. These protein "gates" allow the access of specific biological molecules while excluding other molecules not desired by the cell. In contrast, facilitated diffusion works via passive diffusion of molecules through specific selective transmembrane protein channels [17,18]. These channels, although specific, are largely limited in their control and it is believed that a given nanoparticle of small enough size may be capable of being taken up either by direct transport through the channels or entering the cell with the transport of other larger biological molecules.

Although these mechanisms are common to many molecules, their role in nanoparticle uptake is lesser known: there are evidences that many of these basic transport mechanisms do interact directly with some type of nanoparticles [19]. Although it has not been possible to differentiate the uptake mechanisms in many studies, it is known that many materials, when in nanoparticle form, may enter cells and be readily transported through a number of different mechanisms including endocytosis, active transport, and facilitated diffusion [20–22]. The potential of a nanoparticle to be transported through any one of these mechanisms is inherently and primarily governed by its size, as larger bulk materials or agglomerates have been shown to have poor cellular uptake [23]. However, surface charge and reactivity also determine the likelihood of transport of cellular uptake [23]. Indeed, a number of nanomaterials have been engineered with positively charged surfaces to capitalize on the negatively charged tight junction filaments between cells, increasing their transport through paracellular mechanisms [24]. Furthermore, many nanoparticles have been utilized for their transport properties to enhance the uptake of drugs with poor bioavailability or aqueous solubility [25,26]. Furthermore, nanoparticles in common food products have been found to be readily transported via highly specific cellular mechanisms, such as clathrin-mediated endocytosis in the oral mucosa [19]. Although the transport processes of nanoparticles are likely highly varied dependant on their specific properties, the mechanisms are important to understand their interactions with cells and also their potential to cause harm. Once nanomaterials have entered cells, their potential to cause cytotoxicity or systemic toxicity increases greatly.

5.1.2 Mechanisms of nanoparticle cytotoxicity and systemic toxicity

Throughout the journey of nanomaterials in foods, nanomaterials are exposed to the cells of the GIS that may interact with them in a number of different ways. The way in which a given nanomaterial interacts with these cells depends greatly on the environment of the tissue, the surface charge, size of the nanoparticle, and the type of cell encountered [20]. The field of toxicology accepts that current testing procedures and understanding are not sufficient to truly predict the risks from a wide range of nanoparticles [27]. However, a number of pivotal studies are starting to expose new mechanisms and properties of nanoparticles that clearly demonstrate how nanoscale can drastically change the bioavailability and toxicity of well-known and studied materials. The 2009 paper by Stone et al. highlighted this clearly, presenting data that showed that once nanoparticles are transported into the systemic blood, they can accumulate in many organs of the body [28]. The findings revealed that accumulation readily occurs in the liver, bone, nervous tissues, and even the brain where these nanoparticles could pose major risks to human health [28]. However, the processes by which nanoparticles cause cellular damage can vary drastically: from immune mediated damage to direct effect on membrane integrity. For example, metallic nanoparticles have been shown to cause increased cellular damage compared to larger bulk materials through catalyzing cellular and toxic reactions at a greater rate due to their high surface area [29]. Furthermore, the release of metal ions from these metals as they degrade has been demonstrated to cause cellular damage through oxidative stress [30,31]. This process is caused by metal ions producing free radical species through processes such as the Haber–Weiss cycle and Fenton reaction (Fig. 5.2) [32]. Furthermore, other mechanisms of toxicity indirectly caused by nanomaterials have been discovered that are capable of causing cellular damage through immune system response activation. The activation of immune response to foreign bodies such as metal oxides, carbons, and other organic-based materials is well documented [33–36]. However, due to the nature of nanoparticles and their ability to be transported to a number of different tissues in the body, they pose a risk of the immune response against these materials damaging the function of organs in which they accumulate. Studies on the accumulation of silicon oxide nanoparticles, carbon–iron nanotubes, gold and silver nanoparticles found large build up in the liver and spleen in animal models [37–40]. Although in the case of most of these materials no major toxicity was observed at lower doses, silver nanoparticles however, were identified to cause immunotoxic response [37]. Furthermore, new evidence is emerging linking exposure

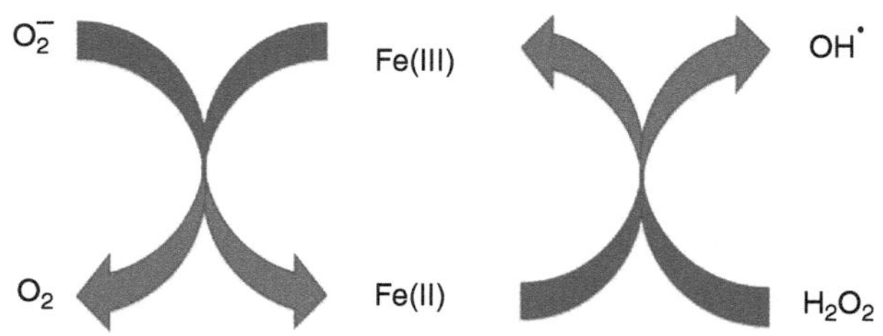

Figure 5.2 Haber–Weiss cycle and Fenton reaction: H_2O_2 produced from mitochondrial activity interacts directly with iron through the Fenton reaction producing reactive hydroxide free radicals that can cause cellular damage by lipid peroxidation.

of nanoparticles to the development of chronic autoimmune diseases such as Crohn's disease through accumulation and activation of immune responses [35].

Immune activation in response to foreign bodies such as metal oxide nanoparticles or carbon nanomaterials, for example, occurs through a process of antibody mediated cellular activation. Antibodies that contact the surfaces of the foreign materials bind and change their structural configuration presenting binding sequences that are recognized by cells of the immune system, in effect, flagging the material to the immune system as foreign. Dependent on the type of cell, the cell can either engulf the foreign body through phagocytosis or can release powerful cytotoxic chemicals into the surrounding tissue. If released in excess, these chemicals can cause damage and inflammation to the host tissue thus reducing its ability to function correctly.

Cytotoxicity of nanoparticles through disruption of intracellular process has also been widely reported with a number of key components and organelles being disrupted, including mitochondria [41], actin cytoskeleton architecture [42], and even the nucleus, resulting in genotoxicity by disruption of DNA by TiO_2 nanoparticles [43].

Although there are undoubtedly a great number of cytotoxic mechanisms currently unknown, those previously discussed show clear mechanisms and evidence that nanoparticles found in foods could pose a potential risk to human health. However, a risk that may arise or be is yet unknown. It inevitably depends greatly on the concentration of nanoparticle administered, on the tissue environment and cells which a nanoparticle encounters. Therefore, to understand bioavailability and toxicity, the environment of the cell must also be understood.

5.2 THE JOURNEY AND BIOLOGICAL EXPOSURE OF NANOPARTICLES IN FOODS

5.2.1 The gastrointestinal system

The GIS is responsible for the breakdown, digestion of foods, the absorption of nutrients, and the removal of waste. The GIS is a varied and complex series of mucosal membranes and compartments each with a very different role and chemical environment, all of which can greatly affect the material dissolution, particle–particle interaction, and adsorption of nanoparticles into the body [44]. The GIS is over 30 feet long and has a total absorptive area of more than 30 m^2, giving it the greatest bioavailability and uptake potential of all systems of the body [45]. This section aims to cover the role and environment of each area of the GIS and how this may affect the bioavailability and behavior of different nanoparticles.

5.2.2 The oral cavity

The oral cavity is a mildly acidic environment with a number of key enzymes that aid the breakdown of foods. It is also home to a number of different cellular types each highly specialized for their role. These cell types typically belong to one of two classes of oral tissues, the keratinized gingival tissue or the nonkeratinized tissue for the buccal mucosa (including cheek, floor of mouth and surface of the tongue and lips, and the mixed mucosa) [46]. The cells of primary contact with nanoparticles are gingival epithelial cells. These cells line the surface of the gingiva (gums) acting as a barrier layer protecting the body from unwanted materials [47]. The gingival tissues are composed of multiple layers of cells that, like skin, regularly shed [47]. The upper layer of the epithelium is formed of highly keratinized epithelial cells making the tissue highly resistant to mechanical force and chemical damage from enzymes while having strong cell to cell tight junctions [47]. Below this layer there are epithelial cells that will later become keratinized and replaced. Below the epithelium, there is a small layer of fibroblastic cells that serve the role of maintaining the tissue composition and to remodel it in the event of damage to the mucosal layer. Beneath the fibroblastic cells, there is a single layer of macrophages that serve to defend the tissue from bacteria viruses and foreign bodies such as nanomaterials. This layer is highly important as the macrophage cells are capable of identifying foreign materials or cells and engulfing and destroying them through phagocytosis [48]. Furthermore, in response to infection or penetration of the tissue by foreign body, they can stimulate a defensive immune response creating an increased influx of immune cells to the tissue as well as an increased permeability of the blood vessels causing gingival bleeding. This process increases the speed of both the breakdown and removal of foreign materials and infective agents to protect the body from unregulated transport into the circulatory system.

As previously discussed the environment of the oral cavity is very mildly acidic with a pH between 6.5 and 6.8 [49]. There are also a number of key enzymes, these include: lysozymes—enzymes that kill bacteria; lactoferrin—anti-viral/bacterial enzyme that also increases the absorption of iron ions into the body [50]; lactoperoxidase—a hydrogen peroxide catalyzing enzyme that aids in the breakdown of both inorganic and organic materials [51]. The influence of this environment and its enzymes on nanoparticles can greatly determine their cytotoxicity, for example, the presence of high concentrations of lysozymes may degrade peptidoglycan or biopolymer nanoparticles such as chitosan used in oral heath care products. Similarly the presence of hydrogen peroxide producing enzymes may accelerate the breakdown of metal-based nanoparticles through reactions such as Fenton's reaction, thus releasing metal ions at an accelerated rate having implications on toxicity [52]. Interestingly, lactoferrin has been shown to have a significant stabilizing effect on nanoparticulate emulsions of curcumin, which is commonly used in nanoparticulate forms as a food additive and colorant [53].

Although the knowledge of nanoparticle uptake in the oral mucosa is still in its infancy, recent research has shown that many types of inorganic nanoparticles have very low or no bioavailability or uptake into the body through the oral mucosa [44]. This is suggested to be a result of multiple factors: a reduced rate of paracellular transport in the keratinized cells has been identified as a potential rational for these findings [44,54]. Secondly, saliva flow, a viscous mucus that forms a tight covering continually flowing away from the oral cavity, is thought to prevent the uptake of nanomaterials [55]. Although inorganic nanoparticles may have a low absorption, the bioavailability of lipid nanoparticles in the oral mucosa has been shown to be evidently greater [56]. The impact of this uptake on human health is currently not clear but the accumulation of nanoparticles of foreign lipids may cause potential immune activation and toxicity [57]. Although bioavailability and uptake into the body is an important determinant of toxicity, nanoparticles can still cause harm even if not transported across a mucosal barrier. This as demonstrated by the work of Best et al. who found ZnO nanoparticulates used in toothpastes were cytotoxic to oral mucosal epithelial cells at concentrations exceeding 0.31% w/v [44]. The long-term implications on oral health from the exposure of nanoparticles in dental products and foods to this extent is not well known. Furthermore, with very limited regulatory restrictions or definitions on the use of these materials in oral exposure products a largely unknown risk exists that must be addressed.

Once passing from the oral cavity, foods have been broken down into small particulates via mastication (chewing) and mild enzymatic digestion. Now, small enough, the foods travel through the esophagus to the stomach.

5.2.3 The esophagus

The esophagus derived from the Greek word *oisophagos*, meaning "to carry to eat," is a tubular tissue that connects the oral cavity and throat to the stomach and is formed of a number of different key layers [58]. Its primary function is to ensure the smooth passage of food from the mouth and throat to the stomach while also preventing the content of the stomach from moving backwards [59]. This is achieved by both muscular action and mucosa secretion to ensure safe transit. As such, the structure of the esophageal tissue, the most inner layer of the lumen that is in contact with food is an epithelial cell layer and mucosa. This serves two processes: firstly the secretion of a mucous that lubricates while the epithelial cells act as a selective barrier to the absorption of nutrients [59]. Below this layer there is a muscular mucosa and a major muscular layer which contracts and relaxes the tissue to create a downward motion to drive the food down. The correct function of this tissue is highly important to maintain correct GIS function; therefore the interaction of nanoparticles is of high importance to food research. Although there are still limited studies of nanoparticle uptake and absorption through the esophagus, it is considered a relatively low area of bioavailability and uptake. This is for two reasons: firstly, the esophagus has relatively small absorptive area compared to the other area of the GIS [59]. Secondly, the duration of contact time between food (or nanoparticles) and the esophagus tissue is typically seconds to minutes whereas other tissues of the GIS, such as the stomach and intestines, are in contact with food materials for hours. However, some studies have found significant toxicity risks of nanomaterials used as contrast agents in magnetic resonance imaging to the esophagus [33]. These studies have shown links between the use of super magnetic iron oxide nanoparticles and the development of tumors and a number of cytotoxicity mechanisms including DNA damage, oxidative stress, changes in gene expression profiles, disturbance in iron homeostasis, and disruption of signaling pathways and impairment of cell cycle regulation [33].

As foods pass from the esophagus to the stomach, they reach the lower section of the esophagus with the lower esophageal sphincter. Here, the sphincter relaxes passing the food into the stomach, where it will be processed by chemical digestion [60].

5.2.4 The stomach

The stomach is a highly acidic environment with a pH typically ranging 1.5–3.5 [59]. It also has a high concentration of proteases enzymes that are released in response to food entering the stomach [59]. These enzymes break down proteins and carbohydrate ready for absorption in the small intestine. Although the stomach also can absorb amino acids and small molecules, most of the absorption and chemical digestion actually occur in the small intestine, specifically, in the duodenum. However, when looking at the factors that affect nanoparticle behavior, the very low pH of the stomach can dramatically affect both the integrity and properties of nanoparticles [61]. Acid erosion of metal-based nanoparticles such as silver can release much higher levels of metal ions than other more neutral pH environments [61]. The release of high levels of metal ions is extremely toxic to cells though the generation of reactive oxygen species leading to oxidative stress [3,62]. Moreover, a range of nanomaterials such as chitosan and nanoemulsions are also readily degraded in the stomach, changing their structural form, size, and integrity [53,61]. As discussed previously, nanoemulsions of curcumin have been stabilized against acid digestion in the stomach using the addition of lactoferrin and enzymes of saliva [53]. This was proposed as a method to reduce the bioavailability of curcumin, which has been shown to be cytotoxic and causing many side effects in clinical trials. Curcumin is of particular concern as a colorant and additive in many foods as it has been found to be an iron chelator, and thus reduces the uptake of nutritional iron [63,64]. Although presenting an improved stability and reduced bioavailability within the stomach, nanoemulsions were less stable when moving to the other parts of the GIS. Most of the nanoemulsion breakdown was observed in the duodenum as this is the main site of chemical digestion. However, the toxicity of these nanomaterials to the stomach is not necessarily related to bioavailability, but to the form they take under certain conditions. For example, the work of Wick et al. demonstrated the effect of the pH of the environment and strong acids on the aggregation of carbon nanotubes [65]. This work documents the reversal of the classical model of bioavailability linked toxicology; the findings revealed a strange paradox by which nanotubes dispersed freely

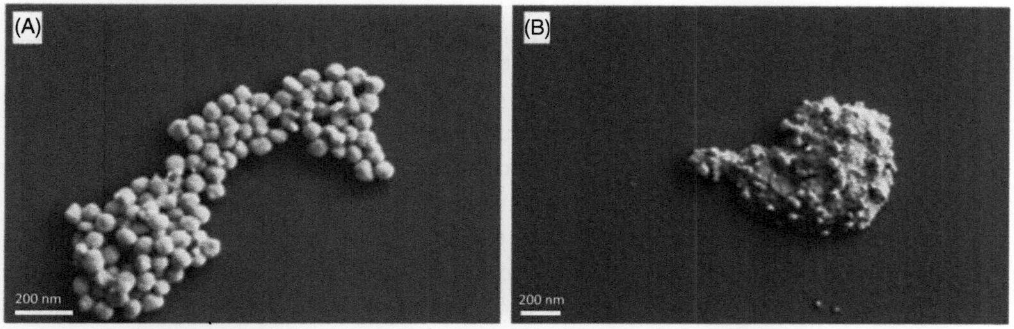

Figure 5.3 Scanning electron microscopy images of silver nanoparticles (stabilized with pvp) after 15 min exposure to synthetic human stomach fluid; (A) 75 nm silver particles and (B) 10 nm silver particles. *(Reproduced from S.K. Mwilu, A.M. El Badawy, K. Bradham, C. Nelson, D. Thomas, K.G. Scheckel, T. Tolaymat, L. Ma, K.R. Rogers, Changes in silver nanoparticles exposed to human synthetic stomach fluid: effects of particle size and surface chemistry, Sci. Total Env. 447 (2013) 90–98 [66] with permission from Elsevier).*

at neutral pH with a high bioavailability showed little toxicity. However, with change in pH, the nanotubes were found to assemble into agglomerates of varying structure. These structures, while having a lower bioavailability, had a greater cytotoxicity [65]. Furthermore, other nanomaterials have also been shown to aggregate under stomach environmental conditions with silver nanoparticles commonly used in foods found to aggregate and thus reducing their bioavailability [66]. Particle size can affect bioavailability; an example is given in Fig. 5.3 where silver nanoparticles of different particle sizes stabilized with polyvinylpyrrolidone (pvp), which have been exposed to media simulating stomach fluid for 15 min. It can be observed that whereas the bigger particles maintain their individual particle shape, the smaller particles coalesce, showing higher degree of transformation and aggregation.

In contrast, for many materials, bioavailability is lower in the stomach. For the case of zinc oxide nanoparticles, the gastric juice of the stomach has been shown to completely dissolve all tested nanoformulations [67], thus demonstrating a high bioavailability through the release of zinc ions and the implications that they have to cytotoxicity. Although this highlights the high range of variability in nanomaterials and how difficult it is to predict their behavior, it also demands that we understand more about the risks and mechanisms of these materials. Furthermore, this research highlights the importance of the influence of strong pH environments and how it can influence on both bioavailability and toxicity to biological systems. From the stomach, digested food (known as chime) is passed through the pyloric sphincter by muscular contraction of the stomach into the duodenum of the small intestine. Any unabsorbed or degraded nanoparticles will pass with this flow of digested food.

5.2.5 The small intestine

The small intestine is formed of three district regions: the duodenum, jejunum, and ileum. The duodenum is where most of the chemical digestion occurs [59]. As discussed previously, the nanoparticle directed autoimmunity has been implicated in Crohn's disease, which affects the endothelium of the duodenum [35]. It is thought that this mechanism is caused by immune activation of macrophage in response to exposure to ultrafine nanoparticles and materials in contact with the epithelium.

Moving from the duodenum and jejunum, chime moves into the ileum, a contractile tissue that controlled the flow of foods based on chemical signaling. The ileum has a pH of 7–8 and is more than 2 m long, having the greatest surface area of any of the tissues of the GIS with more than 20 m^2 [45]. This property has evolved for the purpose of nutrient uptake from foods and, as such, is thought to be the main site of nanoparticle bioavailability and uptake of the GIS. Further to this, as discussed previously, many materials agglomerate in the stomach which reduces their bioavailability [66]. However, it is in the small intestine, specifically in the ileum, where the reduction of pH has shown that silver nanoparticles redisperse, increasing their bioavailability to the ileum epithelium [68].

The ileum epithelium is a highly sensitive tissue and has a great number of nerve collections and a supporting sympathetic nerves system which controls the contraction wave of the ilium that determines the movement rate of the chime. The effects of metals on this tissue have long been known with papers back as far as 1971 showing zinc and cobalt ions interfering with this natural contraction process. These metal ions did not only cause the contraction of the ilium, but also spasm and thus reducing food transit rate and the absorption profile [69]. With zinc nanoparticles so commonly used in foods, and known to dissolve in the acids of the stomach, the generation of zinc ions from everyday foods is of high likelihood and of huge risk to the long-term health of the ileum. Further to this, there are numerous studies documenting the transport and high bioavailability of nanoparticles across the ileum epithelium reaching the circulatory blood [70]. These include titanium dioxide and silver, both present in many food and oral health care products [70]. Based on these findings, the ileum can serve as the main entry point to the body for most nanomaterials in foods, giving them access to the circulatory blood, which as previously discussed, can lead to acumination in many organs and tissues of the body [28]. With the implications of these findings to human health, it is of critical importance that food and drug regulators tighten their requirements over the use of these materials, zinc, silver, and titanium dioxide, in food.

As the ileum removes almost all the remaining nutrients from the chime, the indigestible food (waste) moves into colon where water and remaining nutrients are removed to form the final stool.

5.2.6 The colon

The colon plays an important role in removing large amounts of water and salt from the digested food and, as such, is highly important to maintain water homeostasis in the body [59]. The colon is also home to billions of bacteria that are essential to its correct functioning serving as a major part of the body's natural microflora, with up to 5000 species present in the GIS [71,72]. These bacteria play a critical role in the immunological and digestive function of the GIT. These include the regulation of colonic enterocyte health through the production of short-chain fatty acid butyrate, vitamin B12 and modification of certain drugs and plant toxins reducing their toxicity [72–74]. Furthermore, the microflora of the colon plays an important role in the immune system maturation and education, while also preventing the growth of pathogenic species of bacteria harmful to the body [70,75]. The influence and negative effects of changes to this remarkable environment have been found to have wide ranging implication from cancer risk [76] to xenobiotic metabolism [70,77].

The uptake of nanoparticles in the colon is not well studied but is thought to be less than that of the small intestine. However, due to the important balance of microflora in the colon, the presence of antimicrobial nanoparticles in foods for storage and prevention of spoilage could potentially impact the function of the colon through alteration of the function of the microflora. A recent study by Taylor et al. on the influence of metal oxide nanoparticles on gut microflora phenotype revealed small but significant changes in microflora expression. The authors of the paper proposed this as a possible cause for the development of inflammatory diseases such as ulcerative colitis through changes in expression or in native bacterial population [78]. Other relevant research has found that the epithelial cells of the colon are also highly sensitive to platinum nanoparticles, with cytotoxicity directly related to size of the nanoparticle, and smaller particles increasing cellular death [79]. That research also demonstrated that the toxicity mechanisms were not as expected through metal ion driven free radical generation and oxidative stress, but through other unknown mechanisms of toxicity [79]. These findings demonstrate the huge area of cytotoxicity mechanisms unknown to current knowledge and that many nanoparticles may not only have one mechanism of toxicity but perhaps several. Further to this, any cytotoxic effects on the tissue of the colon can potentially lead to both compromised barrier function and water homeostasis, both critical to many systems of the body. To illustrate the toxicity of platinum nanoparticles, Fig. 5.4 shows a scanning electron microscopy (SEM) image of slices (20 nm thick) of human colon carcinoma cells (so-called HT29) which have interiorized Pt nanoparticles.

The risks of nanoparticles from foods to the health of the colon are already evident from the available literature, with the highlighted links discovered between inflammatory disease in the colon (colitis) and the increase in the use of nanomaterials. It is of critical importance that toxicities of nanomaterials with the colon and the small intestine are areas of intense research focus in the future.

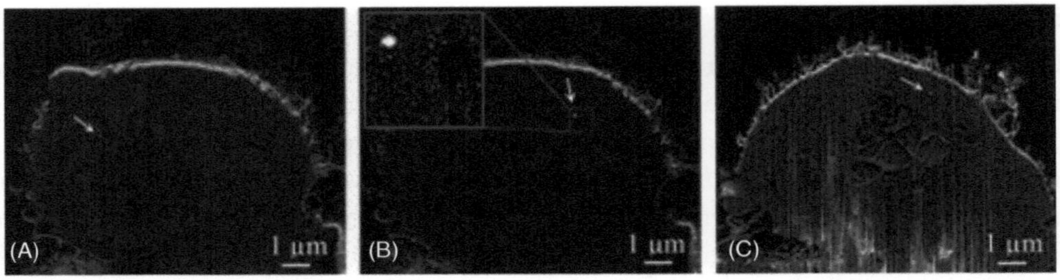

Figure 5.4 Detection of platinum particles in the interior of HT29 cells after incubation (<100 nm particles at a concentration of 1000 ng/cm^2 for 24 h) (A–C) Images of different sections of the HT29 cell arrows mark small platinum clusters. *(Reprinted with permission from J. Pelka, H. Gehrke, M. Esselen, M. Türk, M. Crone, S. Bräse, T. Muller, H. Blank, W. Send, V. Zibat, P. Brenner, R. Schneider, D. Gerthsen, D. Marko, Cellular uptake of platinum nanoparticles in human colon carcinoma cells and their impact on cellular redox systems and DNA integrity, Chem. Res. Toxicol. 22 (4) (2009) 649–659 [79]. Copyright 2009 American Chemical Society).*

5.3 CONCLUSIONS

The global increase seen in inflammatory disease affecting the gastrointestinal tract including ulcerative colitis and Crohn's disease has been linked to the increases in processed foods and the use of nanoparticles in foods [35,80]. Although there is currently insufficient evidence strongly linking these findings, evidence of the risks of cytotoxicity and systemic toxicity posed by nanoparticles is evident. With these present risks identified in current food production and with the continued development of nanoparticle technologies, it is the role of regulatory bodies to ensure the protection of the general public. It is widely accepted in the scientific field that the current methods of analysis of toxicity are not sufficient to understand all the risks posed by nanoparticles in foods. Therefore, a dynamic change is needed to ensure the continued safety of foods.

The advent of new nanoparticle technology has yielded a huge benefit to food science and holds much promise for many future developments. However, the role of food and life scientists in this developing industry is paramount to understanding more about these materials and how they can interact with the human body to understand the risks they pose to human health.

REFERENCES

[1] K. Martin, The chemistry of silica and its potential health benefits, J. Nutr. Health Aging 11 (2007) 94–98.
[2] J.J. Powell, N. Faria, E. Thomas-McKay, L.C. Pele, Origin and fate of dietary nanoparticles and microparticles in the gastrointestinal tract, J. Autoimmun. 34 (3) (2010) J226–J233.
[3] A. Nel, T. Xia, L. Madler, N. Li, Toxic potential of materials at the nanolevel, Science 311 (5761) (2006) 622–627.
[4] G. Oberdorster, A. Maynard, K. Donaldson, V. Castranova, J. Fitzpatrick, K. Ausman, J. Carter, B. Karn, W. Kreyling, D. Lai, S. Olin, N. Monteiro-Riviere, D. Warheit, H. Yang, Principles for characterizing the

potential human health effects from exposure to nanomaterials: elements of a screening strategy, Part. Fibre Toxicol. 2 (8) (2005) 1–35.

[5] W. Forth, D. Henschler, W. Rummel, Pharmacology and toxicology. For students of medicine, veterinary medicine, pharmacy, chemistry, biology as well as for physicians, veterinarians and pharmacists, 1987.

[6] N.P. Innes, G.R. Ogden, A technique for the study of endocytosis in human oral epithelial cells, Arch. Oral Biol. 44 (1999) 519–523.

[7] T. Jung, W. Kamm, A. Breitenbach, E. Kaiserling, J.X. Xiao, T. Kissel, Biodegradable nanoparticles for oral delivery of peptides: is there a role for polymers to affect mucosal uptake? Eur. J. Pharm. Biopharm. 50 (2000) 147–160.

[8] V. Kodali, M.H. Littke, S.C. Tilton, J.G. Teeguarden, L. Shi, C.W. Frevert, W. Wang, J.G. Pounds, B.D. Thrall, Dysregulation of macrophage activation profiles by engineered nanoparticles, ACS Nano 7 (2013) 6997–7010.

[9] R. Merry, L. Belfield, P. McArdle, A. McLennan, S. Crean, A. Foey, Oral health and pathology: a macrophage account, Br. J. Oral Maxillofac. Surg. 50 (2012) 2–7.

[10] C. Buono, J.J. Anzinger, M. Amar, H.S. Kruth, Fluorescent pegylated nanoparticles demonstrate fluid-phase pinocytosis by macrophages in mouse atherosclerotic lesions, J. Clin. Invest. 119 (2009) 1373–1381.

[11] L. Treuel, X. Jiang, G.U. Nienhaus, New views on cellular uptake and trafficking of manufactured nanoparticles, J.R. Soc. Interface 10 (2013) 20120939.

[12] T. Lund, M.F. Callaghan, P. Williams, M. Turmaine, C. Bachmann, T. Rademacher, I.M. Roitt, R. Bayford, The influence of ligand organization on the rate of uptake of gold nanoparticles by colorectal cancer cells, Biomaterials 32 (2011) 9776–9784.

[13] R. Shukla, V. Bansal, M. Chaudhary, A. Basu, R.R. Bhonde, M. Sastry, Biocompatibility of gold nanoparticles and their endocytotic fate inside the cellular compartment: a microscopic overview, Langmuir 21 (2005) 10644–10654.

[14] J.P. Ryman-Rasmussen, J.E. Riviere, N.A. Monteiro-Riviere, Surface coatings determine cytotoxicity and irritation potential of quantum dot nanoparticles in epidermal keratinocytes, J. Invest. Dermatol. 127 (2007) 143–153.

[15] Y. Cai, Y. Liu, W. Yan, Q. Hu, J. Tao, M. Zhang, Z. Shi, R. Tang, Role of hydroxyapatite nanoparticle size in bone cell proliferation, J. Mater. Chem. 17 (2007) 3780–3787.

[16] G.M. Cooper, R.E. Hausman, The Cell: A Molecular Approach, fifth ed., ASM Press, Sunderland, MA, 2009.

[17] M. Friedman, Principles and Models of Biological Transport, second ed., Springer-Verlag, New York, Inc., 2008.

[18] D. Voet, J.G. Voet, C.V. Pratt, Fundamentals of Biochemistry Upgrade, Upgrade edition, John Wiley & Sons, Chichester, 2002.

[19] E. Frohlich, E. Roblegg, Models for oral uptake of nanoparticles in consumer products, Toxicol 291 (2012) 10–17.

[20] A. Asati, S. Santra, C. Kaittanis, J.M. Perez, Surface-charge-dependent cell localization and cytotoxicity of cerium oxide nanoparticles, ACS Nano 4 (2010) 5321–5331.

[21] E.C. Cho, L. Au, Q. Zhang, Y. Xia, The effects of size, shape, and surface functional group of gold nanostructures on their adsorption and internalization by cells, Small 6 (2010) 517–522.

[22] A.S. Holpuch, G.J. Hummel, M. Tong, G.A. Seghi, P. Pei, P. Ma, R.J. Mumper, S.R. Mallery, Nanoparticles for local drug delivery to the oral mucosa: proof of principle studies, Pharm. Res. 27 (2010) 1224–1236.

[23] A. Elder, S. Vidyasagar, L. DeLouise, Physicochemical factors that affect metal and metal oxide nanoparticle passage across epithelial barriers, Wiley Interdiscip. Rev. Nanomed. Nanobiotechnol. 1 (2009) 434–450.

[24] I.M. van der Lubben, J.C. Verhoef, G. Borchard, H.E. Junginger, Chitosan and its derivatives in mucosal drug and vaccine delivery, Eur. J. Pharm. Sci. 14 (2001) 201–207.

[25] J.W. Card, D.C. Zeldin, J.C. Bonner, E.R. Nestmann, Pulmonary applications and toxicity of engineered nanoparticles, Am. J. Physiol. Lung Cell. Mol. Physiol. 295 (2008) L400–L411.

[26] H. Sarin, A.S. Kanevsky, H. Wu, K.R. Brimacombe, S.H. Fung, A.A. Sousa, S. Auh, C.M. Wilson, K. Sharma, M.A. Aronova, R.D. Leapman, G.L. Griffiths, M.D. Hall, Effective transvascular delivery of nanoparticles across the blood-brain tumor barrier into malignant glioma cells, J. Transl. Med. 6 (80) (2008) 1–15.

[27] T.F. Stone, V. Nowack, B. Baun, A. van den Brink, N. von der Kammer, F. Dusinska, M. Handy, R. Hankin, S. Hassellöv, M. Joner, E. Fernandes, Nanomaterials for environmental studies: classification, reference material issues, and strategies for physico-chemical characterisation, Sci. Total Environ. 408 (2010) 1745–1754.

[28] W.G. Kreyling, M. Semmler-Behnke, J. Seitz, W. Scymczak, A. Wenk, P. Mayer, S. Takenaka, G. Oberdörster, Size dependence of the translocation of inhaled iridium and carbon nanoparticle aggregates from the lung of rats to the blood and secondary target organs, Inhal. Toxicol. 21 (2009) 55–60.

[29] A. Kroll, M.H. Pillukat, D. Hahn, J. Schnekenburger, Current in vitro methods in nanoparticle risk assessment: limitations and challenges, Eur. J. Pharm. Biopharm. 72 (2009) 370–377.

[30] C.A. Dick, D.M. Brown, K. Donaldson, V. Stone, The role of free radicals in the toxic and inflammatory effects of four different ultrafine particle types, Inhal. Toxicol. 15 (2003) 39–52.

[31] S. Barillet, A. Simon-Deckers, N. Herlin-Boime, M. Mayne-L'Hermite, C. Reynaud, D. Cassio, B. Gouget, B. Carrière, Toxicological consequences of TiO_2, SiC nanoparticles and multi-walled carbon nanotubes exposure in several mammalian cell types: an in vitro study, J. Nanopart. Res. 12 (2010) 61–73.

[32] C. Buzea, I.I. Pacheco, K. Robbie, Nanomaterials and nanoparticles: sources and toxicity, Biointerphases 2 (2007) MR17–MR71.

[33] N. Singh, G.J. Jenkins, R. Asadi, S.H. Doak, Potential toxicity of superparamagnetic iron oxide nanoparticles (SPION), Nano Rev. 1 (2010) 1–15.

[34] J. Timbrell, Principles of Biochemical Toxicology, Taylor & Francis Ltd., Abingdon, 1991.

[35] M.C. Lomer, R.P. Thompson, J.J. Powell, Fine and ultrafine particles of the diet: influence on the mucosal immune response and association with Crohn's disease, Proc. Nutr. Soc. 61 (2002) 123–130.

[36] T. Yoshida, Y. Yoshioka, H. Takahashi, K. Misato, T. Mori, T. Hirai, K. Nagano, Y. Abe, Y. Mukai, H. Kamada, S.-I. Tsunoda, H. Nabeshi, T. Yoshikawa, K. Higashisaka, Y. Tsutsumi, Intestinal absorption and biological effects of orally administered amorphous silica particles, Nanoscale Res. Lett. 9 (532) (2014) 1–7.

[37] W.H. De Jong, L.T. Van Der Ven, A. Sleijffers, M.V. Park, E.H. Jansen, H. Van Loveren, R.J. Vandebriel, Systemic and immunotoxicity of silver nanoparticles in an intravenous 28 days repeated dose toxicity study in rats, Biomaterials 34 (2013) 8333–8343.

[38] S.K. Balasubramanian, J. Jittiwat, J. Manikandan, C.N. Ong, L.E. Yu, W.Y. Ong, Biodistribution of gold nanoparticles and gene expression changes in the liver and spleen after intravenous administration in rats, Biomaterials 31 (8) (2010) 2034–2042.

[39] T. Hasezaki, K. Isoda, M. Kondoh, Y. Tsutsumi, K. Yagi, Hepatotoxicity of silica nanoparticles with a diameter of 100 nm, Die Pharm. Int. J. Pharm. Sci. 66 (2011) 698–703.

[40] H. Wu, G. Liu, Y. Zhuang, D. Wu, H. Zhang, H. Yang, H. Hu, S. Yang, The behavior after intravenous injection in mice of multiwalled carbon nanotube/Fe_3O_4 hybrid MRI contrast agents, Biomaterials 32 (2011) 4867–4876.

[41] S.M. Hussain, K.L. Hess, J.M. Gearhart, K.T. Geiss, J.J. Schlager, In vitro toxicity of nanoparticles in BRL 3A rat liver cells, Toxicol. In Vitro 19 (7) (2005) 975–983.

[42] S.J.H. Soenen, E. Illyes, D. Vercauteren, K. Braeckmans, Z. Majer, S.C. De Smedt, M. De Cuyper, The role of nanoparticle concentration-dependent induction of cellular stress in the internalization of non-toxic cationic magnetoliposomes, Biomaterials 30 (36) (2009) 6803–6813.

[43] R.Y. Prasad, K. Wallace, K.M. Daniel, A.H. Tennant, R.M. Zucker, J. Strickland, K. Dreher, A.D. Kligerman, C.F. Blackman, D.M. DeMarini, Effect of treatment media on the agglomeration of titanium dioxide nanoparticles: impact on genotoxicity, cellular interaction, and cell cycle, ACS Nano 7 (2013) 1929–1942.

[44] M. Best, G. Phillips, C. Fowler, J. Rowland, J. Elsom, Characterisation and cytotoxic screening of metal oxide nanoparticles putative of interest to oral healthcare formulations in non-keratinised human oral mucosa cells in vitro, Toxicol In Vitro 25 (30) (2015) 402–411.

[45] H. Helander, L. Fändriks, Surface area of the digestive tract—revisited, Scand. J. Gastroenterol. 49 (6) (2014) 681–689.
[46] A.H. Dale, B.A. Salonen, J. Jones, New approaches and concepts in the study of differentiation of oral epithelia, Crit. Rev. Oral Biol. Med. 1 (1990) 167–190.
[47] B.A. Presland, R.B. Dale, Epithelial structural proteins of the skin and oral cavity: function in health and disease, Crit. Rev. Oral Biol. Med. (11) (2000) 383–408.
[48] J.R. Kimball, W. Nittayananta, M. Klausner, W.O. Chung, B.A. Dale, Antimicrobial barrier of an in vitro oral epithelial model, Arch. Oral Biol. 51 (9) (2006) 775–783.
[49] D.J. Aframian, T. Davidowitz, R. Benoliel, The distribution of oral mucosal pH values in healthy saliva secretors, Oral Dis. 12 (4) (2006) 420–423.
[50] L. Sánchez, M. Calvo, J.H. Brock, Biological role of lactoferrin, Arch. Dis. Child 67 (5) (1992) 657–661.
[51] J.H. Kohler, H. Jenzer, Interaction of lactoperoxidase with hydrogen peroxide. Formation of enzyme intermediates and generation of free radicals, Free Radic. Biol. Med. 6 (3) (1989) 323–339.
[52] T.K. Das, M.R. Wati, K. Fatima-Shad, Oxidative stress gated by Fenton and Haber Weiss reactions and its association with Alzheimer's disease, Arch. Neurosci. 2 (2) (2015) 1–8 e20078.
[53] A.C. Pinheiro, M.A. Coimbra, A.A. Vicente, In vitro behaviour of curcumin nanoemulsions stabilized by biopolymer emulsifiers—Effect of interfacial composition, Food Hydrocoll. 52 (2016) 460–467.
[54] M. Garrod, D. Chidgey, Desmosome structure, composition and function, Biochim. Biophys. Acta 1178 (2008) 572–587.
[55] U. Lendenmann, J. Grogan, F.G. Oppenheim, Saliva and dental pellicle—a review, Adv. Dent. Res 14 (14) (2000) 22–28.
[56] S. Kotta, A.W. Khan, K. Pramod, S.H. Ansari, R.K. Sharma, Exploring oral nanoemulsions for bioavailability enhancement of poorly water-soluble drugs, Expert Opin. Drug Deliv. 9 (5) (2012) 585–598.
[57] D.J. McClements, Edible lipid nanoparticles: digestion, absorption, and potential toxicity, Prog. Lipid Res. 52 (4) (2013) 409–423.
[58] D. Harper, Oesophagus, Etymology Online, 2014.
[59] H. Gray, Anatomy of the Human Body, 20th ed., Bartleby.Com, New York, 1918.
[60] J.E. Hall, Guyton and Hall Textbook of Medical Physiology, 13th ed., Saunders, Philadelphia, 2016.
[61] R.Y. Yada, N. Buck, R. Canady, C. DeMerlis, T. Duncan, G. Janer, L. Juneja, M. Lin, D.J. McClements, G. Noonan, J. Oxley, C. Sabliov, L. Tsytsikova, S. Vázquez-Campos, J. Yourick, Q. Zhong, S. Thurmond, Engineered nanoscale food ingredients: evaluation of current knowledge on material characteristics relevant to uptake from the gastrointestinal tract, Compr. Rev. Food Sci. Food Saf. 13 (4) (2014) 730–744.
[62] Q. Zhang, Y. Kusaka, K. Sato, K. Nakakuki, N. Kohyama, K. Donaldson, Differences in the extent of inflammation caused by intratracheal exposure to three ultrafine metals: role of free radicals, J. Toxicol. Environ. Health A 53 (1998) 423–438.
[63] A.L. Cheng, C.H. Hsu, J.K. Lin, M.M. Hsu, Y.F. Ho, T.S. Shen, J.Y. Ko, J.T. Lin, B.R. Lin, W. Ming-Shiang, H.S. Yu, S.H. Jee, G.S. Chen, T.M. Chen, C.A. Chen, M.K. Lai, Y.S. Pu, M.H. Pan, Y.J. Wang, C.C. Tsai, C.Y. Hsieh, Phase I clinical trial of curcumin, a chemopreventive agent, in patients with high-risk or pre-malignant lesions, Anticancer Res. 21 (4b) (2001) 2895–2900.
[64] Y. Jiao, J. Wilkinson 4th, X. Di, W. Wang, H. Hatcher, N.D. Kock, R. D'Agostino Jr., M.A. Knovich, F.M. Torti, S.V. Torti, Curcumin, a cancer chemopreventive and chemotherapeutic agent, is a biologically active iron chelator, Blood 113 (2) (2009) 462–469.
[65] P. Wick, P. Manser, L.K. Limbach, U. Dettlaff-Weglikowska, F. Krumeich, S. Roth, W.J. Stark, A. Bruinink, The degree and kind of agglomeration affect carbon nanotube cytotoxicity, Toxicol. Lett. 168 (2) (2007) 121–131.
[66] S.K. Mwilu, A.M. El Badawy, K. Bradham, C. Nelson, D. Thomas, K.G. Scheckel, T. Tolaymat, L. Ma, K.R. Rogers, Changes in silver nanoparticles exposed to human synthetic stomach fluid: effects of particle size and surface chemistry, Sci. Total Environ. 447 (2013) 90–98.
[67] S.H. Seok, W.S. Cho, J.S. Park, Y. Na, A. Jang, H. Kim, Y. Cho, T. Kim, J.R. You, S. Ko, B.C. Kang, J.K. Lee, J. Jeong, J.H. Che, Rat pancreatitis produced by 13-week administration of zinc oxide nanoparticles: biopersistence of nanoparticles and possible solutions, J. Appl. Toxicol. 33 (10) (2013) 1089–1096.

[68] A.P. Walczak, R. Fokkink, R. Peters, P. Tromp, Z.E. Herrera Rivera, I.M. Rietjens, P.J. Hendriksen, H. Bouwmeester, Behaviour of silver nanoparticles and silver ions in an in vitro human gastrointestinal digestion model, Nanotoxicology 7 (7) (2013) 1198–1210.
[69] H. Schnieden, R. Small, Spasmolytic effects of cadmium and zinc ions upon the guinea-pig isolated ileum preparation, Br. J. Pharmacol. 41 (3) (1971) 488–499.
[70] I.L. Bergin, F.A. Witzmann, Nanoparticle toxicity by the gastrointestinal route: evidence and knowledge gaps, Int. J. Biomed. Nanosci. Nanotechnol. 3 (1-2) (2013).
[71] E. Zoetendal, C. Collier, E. Al, Molecular ecological analysis of the gastrointestinal microbiota: a review, J. Nutr. 134 (2) (2004) 465–472.
[72] J. Manson, M. Rauch, M. Gilmore, The commensal microbiology of the gastrointestinal tract, Adv. Exp. Med. Biol. 635 (2008) 15–28.
[73] K. Mason, G. Huffnagle, M. Noverr, J. Kao, Overview of gut immunology, Adv. Exp. Med. Biol. 635 (2008) 1–14.
[74] V. Young, The intestinal microbiota in health and disease, Curr. Opin. Gastroenterol. 28 (2012) 63–69.
[75] S. Walk, V. Young, Emerging insights into antibiotic-associated diarrhea and clostridium difficile infection through the lens of microbial ecology, Interdiscip. Perspect. Infect. Dis. 125081 (2008) 1–7.
[76] R. Canani, M. Costanzo, L. Leone, M. Pedata, R. Meli, A. Calignano, Potential beneficial effects of butyrate in intestinal and extraintestinal diseases, World J. Gastroenterol. 17 (12) (2011) 1519–1528.
[77] T. Clayton, D. Baker, J. Lindon, J. Everett, J. Nicholson, Pharmacometabonomic identification of a significant host-microbiome metabolic interaction affecting human drug metabolism, Proc. Natl. Acad. Sci. USA 106 (34) (2009) 14728–14733.
[78] A.A. Taylor, I.M. Marcus, R.L. Guysi, S.L. Walker, Metal oxide nanoparticles induce minimal phenotypic changes in a model colon gut microbiota, Environ. Eng. Sci. 32 (7) (2015) 602–612.
[79] J. Pelka, H. Gehrke, M. Esselen, M. Türk, M. Crone, S. Bräse, T. Muller, H. Blank, W. Send, V. Zibat, P. Brenner, R. Schneider, D. Gerthsen, D. Marko, Cellular uptake of platinum nanoparticles in human colon carcinoma cells and their impact on cellular redox systems and DNA integrity, Chem. Res. Toxicol. 22 (4) (2009) 649–659.
[80] N.A. Molodecky, I.S. Soon, D.M. Rabi, W.A. Ghali, M. Ferris, G. Chernoff, E.I. Benchimol, R. Panaccione, S. Ghosh, H.W. Barkema, G.G. Kaplan, Increasing incidence and prevalence of the inflammatory bowel diseases with time, based on systematic review, Gastroenterology 142 (1) (2012) 46–54.

CHAPTER SIX

Microbiological Toxicity of Nanoparticles

Samuel Eduok*,, Frédéric Coulon***
*Cranfield University, Cranfield, United Kingdom
**University of Uyo, Uyo, Nigeria

Contents

6.1 INTRODUCTION

There has been a remarkable growth of nanotechnology and growing interest in the application of engineered nanoparticles (ENPs) in several products over the last decade [1,7]. This growth of interest has been associated with increased concern for human and environmental health due to the potential toxicological implications of engineered nanoparticles released into the environment which could have reversible or irreversible effects on microbial-dependent processes [1]. Despite the great research attention commanded by (ENPs) effect on biological systems in recent years, there is still a considerable challenge in the analytical procedures and evaluation [2–4]. ENPs exert their antimicrobial effect through a wide range of mechanisms including the formation of reactive oxygen species (ROS), disruption of microbial physiology and metabolic processes although there is growing evidence that the ENPs could also augment microbial-mediated processes in the ecosystem [4,6,8].

Interaction mechanisms between nanoparticles and microorganisms are not yet fully understood [1,8]. The complexity comes with the particles ability to bind and interact with biological matter and change their surface characteristics, depending on the environment they are in [1,2]. Recent studies on nanoparticle–cell interaction mechanisms

Emerging Nanotechnologies in Food Science

indicated that cells readily take up nanoparticles via either active or passive mechanisms. In contrast intracellular mechanisms and pathways are more difficult to understand as even particles of the same material can show completely different behavior due to, for example, slight differences in surface coating, charge, or size. This makes the categorization of nanoparticle behavior, when in contact with microbes, intricate and thus nanoparticle toxicity is not straightforward. Therefore the entry route of ENPs and their toxicity need to be understood.

6.2 NANOPARTICLES PROPERTIES AND REACTIVITY

The small size of ENPs confers unique physicochemical properties different from its bulk counterpart, offering a broad spectrum of application in consumer products. ENPs can be of spherical, tubular, or irregular shape, be present in fused or aggregated form and are usually designed for a particular product manufacture [3]. Their novel properties are influenced by their homogeneous nature, size of the particle, shape, surface charge, zeta potential, solubility, aggregation potential, interaction with other compounds, and ligands in the medium. ENPs of 1–10 nm exhibit extraordinary properties in relation to their electronic, magnetic, optical, and catalytic reactions when compared with their atomic/molecular and bulk counterparts [3]. Concepts on the special properties of nanoparticles have been discussed in Chapter 1 of this volume. Alteration in physicochemical properties by adsorption influences the nature of ENPs and the surface charge plays a prominent role in the adsorption process. The unique surface of pristine ENPs could be chemically and biologically functionalized or modified by organic matter and microorganisms. Also, the release of functionalized or embedded ENPs is affected by environmental conditions, for example, light, oxidizing agents, or microbial interactions [4].

6.3 NANOPARTICLES APPLICATIONS AND THE ENVIRONMENT

ENPs are produced through the structural organization of groups of atoms and molecules with one dimension of 1–100 nm. ENPs are materials with unique characteristics and enhanced functionalities which attract considerable interest for research and consumer product application [5]. ENPs can be of spherical, tubular, or irregular shape, be present in fused or aggregated form, and are usually designed for a particular product manufacture. Their unique properties are influenced by their homogeneous nature, size of the particle, shape, surface charge, zeta potential, solubility, aggregation potential, interaction with other contaminants, and ligands in the medium.

ENPs such as silver oxide (Ag^0), zinc oxide (ZnO), titanium dioxide (TiO_2), cerium oxide (CeO_2), and silicon oxide (SiO_2) are widely used in consumer products due to their broad microbiocidal and microbiostatic properties and their selective toxicity to microorganisms [8]. More than 1500 consumer products in the health and fitness sector, pharmaceuticals, food, and textiles personal care products, cosmetics, pesticides, paints, sprays, preservatives, diagnostics, therapeutics, surgical devices, nanomedicine, water and wastewater treatment, and formulations for soil remediation contain ENPs [6,7]. ENPs are known to associate with organic materials in sewage sludge, and it is perceived that the environment will be inundated with aged-ENPs released from consumer products into wastewater treatment plants during sludge disposal and effluents discharges [8,9] However, there are no empirical records of residual ENPs in environmental media which heightens environmental health concern to date [8].

6.4 ENGINEERED NANOPARTICLES AND TOXICITY

6.4.1 ENPs and microbiological indices for evaluation

Currently, there are no accepted protocols for toxicity and exposure assessment of ENPs. However to measure toxicological end points, there is a need to fully understand and characterize the ENPs [4,10]. Otherwise, possible toxic effects cannot be easily attributed to a certain property of the ENPs or even the ENP itself because, for example, impurities and other components could be held responsible. Therefore a number of parameters have been recommended as critical requirements for evaluating ENPs reactivity and toxicity including (1) solid-water partition coefficient; (2) octanol–water partition and affinity coefficient; and (3) reaction rate constants. Such information provides an understanding on how ENPs dissolve, precipitate, and degrade [10]. In addition to these parameters, the measure of primary particle size; surface area; zeta potential; impurities; presence of natural organic matter (NOM); and divalent ions in the medium are required to measure ENPs toxicity [4,10]. The biological end points used in the measurement of toxic effect on microbial species include microbial activities, community structure and dynamics, abundance and diversity, species richness, evenness, redox activity, growth, reproduction, synthesis of secondary metabolites such as exopolysaccharides and biofilm formation, bioluminescence, disruption of cell wall/membrane, enzyme activity, alterations in genetic information, and oxygen uptake rate due to cryptic growth or endogenous respiration which vary for different organisms [11,12].

6.4.2 Causal mechanism of ENPs toxicity

ENPs concentration or dose regime is one of the key parameters needed in nanotoxicology to draw meaningful conclusions from in vitro and in vivo experiments for health and environment risk assessment. Thus, nanotoxicology should be determined using nanoparticle dose based on real-world doses rather than unrealistically high doses to achieve a biological response. Generally, toxicity bioassays directly expose the targeted microbial species to ENPs concentrations in diluent. Prior to exposure, in most studies, the ENPs are pretreated by sonication, filtration, dispersion, and drying [13]. However, the pretreatment processes can reduce or enhance the bioavailable ENPs and ultimately influence microbial response in a static or continuous flow toxicity assay [13]. The major mechanisms by which ENPs exhibit toxic effect on microorganisms include the following: (1) oxidative stress from ROS causing lipid peroxidation and alteration of cell permeability, leaking of intracellular content and death [14]; (2) the release of ions and free radicals to interact with key receptors causing protein and amino acid, DNA and nucleotide oxidation, and strand breakage indicated by disruption of microbial activity, physiology, and enzyme functions [11,15,16].

6.4.3 Physico-chemical and biological characteristics dependent toxicity

6.4.3.1 Particle size/shape influences ENPs toxic effect

A major determining factor of ENPs toxic effect is the particle size. The smaller the diameter of a spherical ENP the more the surface-to-volume ratio increases which is accompanied with an increased chemical reactivity and toxicity to microorganisms (12; Table 6.1 summary).

For example, the cytotoxic effect of Ag^0 nanoparticles with particle size <30 nm on *E. coli* and *S. aureus* [25] is more pronounced than Ag^0 nanoparticles with particle size ranging between 80 and 90 nm [31]. In other study, Ag^0 nanoparticles with particle size <5 nm and in suspension with diverse components in wastewater are toxic to nitrifying bacteria in activated sludge [74]. Apart from particle size and dose response, shape also plays a role in nanotoxicity, for instance, Ag^0 exist in the triangular, spherical, or rod-shaped forms. However, when the effects of these three distinct shapes are compared, the truncated triangular form of Ag^0 exerted more bactericidal effect on *E. coli* in both agar plate and broth cultures [75].

6.4.3.2 Ions mediate ENPs toxic effect

The release of ions by ENPs is a potent mechanism to exert toxic effect on microorganisms in different media. However complexion reactions occurring with organic compounds in complex environmental matrices can mitigate such effects by diluting the ion concentrations [76]. The release of ions is also dependent on the alkalinity and hardness of the medium whereas the toxic effect of the released ions depends on the bacterial cell wall composition [70]. The presence of divalent cations interact to form ion bridge with the bacterial cell wall which exacerbate ENPs toxic effect [8,77]. In addition, ENPs such as ZnO, TiO_2, and SiO_2 are phototoxic due to the formation of ROS and free radicals that are enhanced in the presence of light [16,78,79]. However, comparative studies show that C_{60} fullerenes readily associate with *E. coli* than *B. subtilis* and suggest differences in sorption potential due to surface charges [52]. Similarly, Ag^0 preferentially bind to glycoprotein spikes of HIV-1 viruses [80], although opinions are divergent for role of bacterial cell wall and its charges on different ENPs sorbed to it [8].

Table 6.1 Summary of antimicrobial activity of engineered nanoparticles

ENPs	Size (nm)	Test organism	Effect on microorganism	References
Ag^0 (*Shewanella oneidensis* synthesized)	~2–11	*Escherichia coli, Bacillus subtilis, S. oneidensis*	MIC of 2.0, 0.5, and 3.0 μg/mL for *E. coli, B. subtilis, S. oneidensis*	[17]
Ag^0	3–8	*E. coli, B. subtilis, S. oneidensis*	MIC of 6.0, 2.0, and 6.5 μg/mL, respectively	[17]
Ag^0-β-CD, Ag^0	4–7, 17	*E. coli* ATCC 11229, *Pseudomonas aeruginosa* ATCC 27852, *Staphylococcus aureus* ATCC 25923	Capped Ag^0 exhibited about 3.5-fold higher antibacterial activity than uncapped form	[18]
Ag^0 (*Sesuvium portulacastrum* L. callus and leaf extract synthesized)	5–20	Bacteria (*P. aeruginosa, S. aureus, L. monocytogenes, Micrococcus luteus, Klebsiella pnuemoniae*), Fungi (*Alternaria alternaria, Penicillium italicum, Fusarium equisetti, and Candida albicans*)	Antibacterial and antifungal activity with inhibitory zones of 8–23 mm and 12–18 mm, respectively	[19]
Ag^0	8.4, 16.1, 98	*Streptococcus mutans*	Particle size dependent toxicity with MIC of 101 ± 72.03, 145 ± 104.88, and 320 ± 172.88 μg/mL, respectively	[20]
Ag^0	~9	*E. coli*	Inhibition of $AgNO_3$ preexposed bacteria at 3 μM	[21]
Ag^0	9–12	Nitrifying bacteria	Inhibition of microbial growth at EC_{50} of 0.14 mg/L	[22]
Ag^0	10, 30–40, and ~100	Methicillin resistant *S. aureus*	MIC_{90} and MIC_{99} at 1.35 mg/mL, MBC value inhibited 100% of bacterial growth	[23]
Ag^0	~ 12	*E. coli*	70% inhibition of bacterial activity with 10 μg/mL	[24]
Ag^0	20–25	*E. coli, P. aeruginosa, S. aureus*	MIC of 0.53, 0.37, and 0.74 μg/mL compared with 0.83, 1.33, and 0.42 μg/mL for gentamicin, respectively.	[25]
Ag^0	20–25	*C. albicans*	MIC of 6 μg/mL compared with 64 μg/mL fluconazole, and 0.25 μg/mL amphotericin B	[25]

(*Continued*)

Table 6.1 Summary of antimicrobial activity of engineered nanoparticles (*cont.*)

ENPs	Size (nm)	Test organism	Effect on microorganism	References
Ag^0	20–25	*Mycobacterium smegmatis*	MIC of 0.46 μg/mL compared with 0.85 μg/mL rifampicin	[25]
Ag^0	20–25	*Cryptococcus neoformans, C.albicans, Aspergillus niger*	MIC of 3, 6, and 25 μg/mL, respectively	[25]
Ag^0 (*Aspergillus clavatus* synthesized)	20–25	*E. coli, Pseudomonas fluorescens, C. albicans*	Bacterial MIC of 5.83 μg/mL and fungal MIC of 9.7 μg/mL	[26]
Ag^0 (*Streptomyces hygroscopicus* synthesized)	20–30	*B. subtilis, Enterococcus faecalis, E. coli, Salmonella typhimurium, C. albicans*	Broad spectrum growth inhibition of test organisms at 1% v/v	[27]
Ag^0 (*Acalypha indica* leaf extract synthesized)	20–30	*E. coli, Vibrio cholerae*	MIC of 10 μg/mL	[28]
Ag^0-cellulose	20–45	*E. coli, S. aureus*	MIC of 2.7 and 5.4 μg/mL, respectively	[29]
Ag^0-TiO_2	250–300	*C. neoformans, C.albicans, A. niger*	MIC of 3.1, 6, and 12.5 μg/mL respectively	[25]
Ag^0-TiO_2	250–300	*Uropathogenic E. coli*	MIC of 0.24 μg/mL	[25]
Ag^0-TiO_2	26–56	*Nonpathogenic E. coli BL-21*	MIC of 25.46 μg/cm^2	[30]
Ag^0	15–21	*E. coli* PHL628-gfp	MBC of 38 and 10 mg/L killing 99.9% of planktonic and biofilm bacterial population	[22]
Ag^0	7, 29, and 89	*E. coli* ATCC 25922, *S. aureus* ATCC 25923	MIC of 6.25, 13.02, and 11.79 μg/mL, and 7.5, 16.67, and 33.71 for the respective sizes	[31]
Ag^0 (Myramistin stabilized)	10 ± 1.8	*E. coli* ATCC 25922, *S. aureus* FDA 209P (St. 209) *strain, Leuconostoc mesenteriodes* VKPM B-4177, *Saccharomyces cerevisiae* RIA 259, *and A. niger*	MIC of < 1 μg/mL (*E. coli*) and 5 μg/mL (*S. aureus*) on liquid medium, 2.5 μg/L on both liquid and agar plates. 5 μg/mL for *L. mesenteriodes, S. cerevisiae,* and *A. niger*	[32]
Ag^0	13.83	*E. coli* (ATCC 700926), *B. subtilis* (ATCC 9372), *S. cerevisiae* (ATCC 26108)	100% inhibition of *E. coli* at 1.589 μg/mL after 1 h, two orders of magnitude lower effect on *B. subtilis* than *E. coli*, no observable inhibitory effect on *S. cerevisiae*	[33]

Ag^0	100	Estuarine sediment bacterial consortia	No observable effect on bacterial density and genetic diversity after 30 days exposure at 25 μg/L or 1000 μg/L	[34]
Ag^0-TiO_2	<5	*E. coli*	98% growth inhibition at 1.2 μg Ag^0/mL, 100% inhibition at 2.4 μg Ag^0/mL whereas < 3 nm Ag/TiO_2 of 3.9 mg/L produced 99.9% inhibition	[35]
Ag^0	10–20	*E. coli, S. aureus*	MIC of 209 μg/mL for *E. coli* and 433 μg/mL for *S. aureus*	[36]
Ag^0	100	Multidrug-resistant *P. aeruginosa (MRPA),* ampicillin-resistant *E. coli* 0157:H7 (AREC), erythromycin-resistant *Streptococcus pyogenes* (ERSP)	MIC of 66.7 ± 17.7 mM for ERSP, 83.3 ± 16.7 for AREC and MRPA. 99.7, 95.7, and 92.8% MBC at 50 mM to ERSP, AREC, and MRPA, respectively.	[37]
Ag^0	5	*E. coli* ATCC 8739	Complete inhibition of growth and viability at 10 μg/mL for 10^7 cfu/mL	[11]
Ag^0-TiO_2	100	*E. coli WP2 trp uvrA*	Complete photoactivated inhibition of growth and viability at 2.0 mg/L. TiO_2 alone had no observable effect	[38]
Ag^0 (JS47N and SO-01)	20–50	*Microcystis aeruginosa* (cyanobacterium)	Selective reduction in composition ratio of *M. aeruginosa* from 95.5% to 49% and 21% after 10 days incubation at 1 mg/L. Growth inhibition by 87%	[39]
Ag^0 (Hydrophilic)	~7	*E. coli, Shigella sonnei, Bacillus megaterium, Proteus vulgaris, S. aureus*	MIC of 100 μg/mL (*E. coli*), 215 μg/mL (*S. sonnei*), 275 μg/mL (*P. vulgaris*), 300 μg/mL (*B. megaterium*), 350 μg/mL (*S. aureus*)	[40]
Ag^0 (*Bacillus licheniformis* synthesized)	50	*P. aeruginosa, S. epidermidis*	95 and 98% disruption in extracellular polymeric substance/ biofilm formation at 24 h exposure to 100 nM whereas 50 nM resulted in 50% reduction. 12 ± 1.2 mm (*S. epidermidis*) and 9.5 ± 0.9 mm (*P. aeruginosa*) zones of inhibition	[41]

(Continued)

Table 6.1 Summary of antimicrobial activity of engineered nanoparticles (*cont.*)

ENPs	Size (nm)	Test organism	Effect on microorganism	References
Ag^0 (*A. niger* synthesized)	3–30	*E. coli, S. aureus, A. niger, Bacillus species*	Growth inhibition with zone diameters of 12 mm (*A. niger*), 9 mm (*S. aureus*), 8 mm (*Bacillus* sp and *E. coli*)	[42]
Ag^0 (Banana peel extract synthesized)	100	*C. albicans* (BX and BH), *C. lipolytica* (NCIM 3589); *Citrobacter koseri* (MTCC 1657), *Enterobacter aerogenes* (MTCC 111), *E. coli* (MTCC 728), *Klebsiella* sp, *P. vulgaris* (MTCC 426) *P. aeruginosa* (MTCC 728)	Inhibitory zones for *C. albicans* BX (12 mm), *C. albicans BH* (13 mm), *E. aerogenes* (13 mm), *E. coli* (14 mm), *Klebsiella* sp (17 mm); no inhibitory effect on *C. koseri, P. vulgaris, P. aeruginosa*	[43]
Ag^0, CuO, ZnO	10, 33, and 50–70	*Pseudomonas putida KT2440*	Dose-dependent bacteriostatic effect on light output at > 0.2 $mgAg^0/L$, 10 mgCuO/L, < 7 mgZnO/L, whereas lower concentrations of ZnO had stimulatory effect on light production	[44]
Ag^0- *Tribulus terrestris* synthesized	16–28	Mutidrug resistant *S. aureus, B. subtilis, E. coli, P. aeruginosa, S. pyogenes*	High bactericidal activity with inhibition zone range of 9.25– 10.75 mm	[45]
Ag^0 (citrate capped)	10	*E. coli* KACC10495, *B. subtils* KACC10111	Colony forming rates of 0.5% (*E. coli*) and 77.5% (*B. subtilis*) at 10 mg/L	[46]
Ag^0-Graphene (GrO) hybrid	5–25	*E. coli, P. aeruginosa,*	Concentration-dependent growth inhibition of different Ag^0-GrO hybrids	[47]
Ag^0	15–21	Autotrophic nitrifying bacteria	More than 80% growth inhibition by 1 mg/L	[22]
Au, Ag^0, Fe_3O_4	10, 2, and 7, respectively	*Photobacterium phosphoreum*	No inhibition of luminescence at 28, 45, and 52 μg/mL, respectively	[48]
AL_2O_2	<50	*E. coli WP2 trp uvrA,*	No mutagenic effect or growth inhibition on test organism	[38]
Alumina	179	*E. coli,*	Mild to moderate inhibition of microbial growth at a concentration of 10–1000 μg/mL	[49]

Boron	10–20	*Vibrio fischeri* NRRL B-11177	Toxic with EC_{50} of 55.85–65.98 mg/L at 1 and 60 days of the solution age, respectively	[50]
CeO_2	6.5	*V. fischeri*	More than 80% inhibition of luminescence at MIC of 0.064 mg/mL	[51]
CeO_2	<25	Cyanobacteria *Anabaena* CPB43337	24 h EC_{50} of 0.27–6.3 mg/L with menbrane disruption and cell damage	[15]
CO_3O_4	<50	*S. typhimurium* TA 97a and TA100, *E. coli WP2 trp uvrA*	No mutagenic potential or growth inhibition on test organism at tested concentrations	[38]
C_{60} Fullerenes	Variable and dependent on medium	*E. coli, B. subtilis*	MIC of 0.5–1.0 mg/L and 1.5–3.0 mg/L, respectively	[52]
CuO	<50	*S. typhimurium TA 97*a *and TA100*	Low mutagenic potential to test organism at specific concentrations	[38]
CuO	~30	*V. fischeri*	Toxic with 72 h EC_{50} value of 79 mg/L, MIC value of 200 mg/L	[53]
Cu-doped TiO_2	20	*M. smegmatis, S. oneidensis MR-1*	High toxicity level compared with no observable effect at < 100 mg/L of TiO_2	[54]
Fe_3O_4, Au	8, 5	*E. coli*	Dose-dependent microbiostatic effect by Fe_3O_4 (50–200 µg/mL); no observable effect by Au (25–100 µg/mL)	[55]
Fe_3O_4	6	*V. fischeri*	EC_{50} of 0.24 mg/mL	[51]
Fe and Cu	25, 25	*Trametes versicolor*	Significant reduction of lignocellulolytic enzymes; β-glucosidase, β-xylosidase, and cellobiohydrolase at 0.1 mmol/L after 24 h. Reduction in laccase production by copper nanoparticle. Production profile of enzymes not growth related	[56]
TiO_2 (81% anatase, 19% rutile) and ZnO (100% zincite)	15–20 and 20–30	Soil microbial consortia	Altered microbial diversity, low DNA and changes in bacterial composition at 0.5 mg/g soil after 60 days exposure	[12]

(Continued)

Table 6.1 Summary of antimicrobial activity of engineered nanoparticles (*cont.*)

ENPs	Size (nm)	Test organism	Effect on microorganism	References
TiO_2	15–30	*E. coli, P. aeruginosa*	Particle size dependent bactericidal effect on test organisms	[57]
TiO_2, SiO_2, and ZnO	330, 205, and 480	*E. coli, B. substilis*	Photo-induced bactericidal activities on *B.subtilis* (75%) and *E. coli* (44%), SiO_2 least toxic	[58]
TiO_2	< 100	*V. fischeri*	No toxic effect at a concentration of 1000 mg/L	[59]
TiO_2, Au-capped TiO_2	12–18, 5–10	*E. coli* DH 5α, *B. megaterium* QM B1551	60%–100% killing efficiency by photocatalytic formation of reactive oxygen species (ROS)	[60]
TiO_2	7.5	*V. fischeri*	Only 21% inhibition of luminescence detected at a maximum concentration of 1.12 mg/mL	[51]
TiO_2 and ZnO	30 and 50	*S. typhimurium* strain TA98, TA1537 and *E. coli* (WP2 uvrA)	Concentration-dependent uptake and internalization of nanoparticles with 8 and 80 ng/mL after 60 min. Exhibition of weak mutagenic effect.	[61]
TiO_2	25–70	*V. fischeri*	Not acutely toxic: EC_{50} >20 g/L even at 8 h exposure in the dark, MIC: >20 g/L	[53]
TiO_2, CuO, ZnO and Ag^0, fullerene	25, 30, 70, and <100 respectively	Wild type *E. coli* AB1157 (pSLlux), *E. coli* sodABC strain AS391 (pSLlux), *E. coli* K12:soxRSsodAlux	TiO_2 inhibited viability at > 4,000 mg/L, but stimulated luminescence at 100 mg/L. 2 h EC_{50} of 8.1 and 2.0 mg/L (CuO), 46 and 3.1 mg/L (Ag^0), 4.5 and 54 mg/L (ZnO) for triple *sod* mutant and wild type, respectively. Fullerene inhibited bioluminescence of *sod* triple mutant at 3,882 mg/L but NOEC on wild type at 20,800 mg/L	[62]
TiO_2	<100	*E. coli WP2 trp uvrA*	Induced marginal mutagenesis to organism	[38]

TiO_2, ZnO, CuO	25–70, 50–70, and 30, respectively	*S. cerevisiae*	TiO_2 not toxic at EC_{50} of > 20 g/L; 8 and 24 h EC_{50} of 20.7 mg CuO/L, 13.4 mg CuO/L and 121–134 mg ZnO/L and 131–158 mg ZnO/L, respectively	[63]
TiO_2	79	*E. coli* sodABC	Phototoxic effect by reactive oxygen species	[64]
TiO_2	6	*V. fischeri* NRRL B-11177	EC_{50} range of 56–66 mg/L was dependent on the age of TiO_2 solution	[50]
TiO_2	6–40	*V. fischeri*	14% loss of bioluminescence, low toxicity in aqueous medium, LOEC of 500–1000 mg/L (EC_{50} of 650.6 and 940.6 mg/L)	[33]
TiO_2, Al_2O_3, ZnO, SiO_2	50, 60, 20, and 20, respectively	*B. subtilis, E. coli, P. fluorescens*	Toxicity at 20 mg/L: ZnO (100% toxic effect on all organisms), Al_2O_3 (57, 36, and 70%, respectively), SiO_2 (40, 58, and 70%, respectively), TiO_2 (low toxic effect).	[16]
Zero-valent iron nanoparticles	20–30	*B. subtilis var niger, P. fluorescens*	Complete bacterial inactivation at 10 mg/mL, 95 and 80% inactivation at 0.1 and 1 mg/mL, respectively, for *B. subtilis var niger,* complete inactivation of *P. fluorescens* at both concentrations	[65]
ZnO	20	*S. oneidensis* MR-1, *E. coli*	No growth inhibition at > 40 mg/L in aquatic media whereas aerosolized particle of 480 nm and 20 nm was toxic	[54]
ZnO	20–45	*S. aureus, E. coli*	Enhanced antibacterial effect of ciprofloxacin (27 and 22% increased inhibition zones for *S. aureus* and *E. coli*, respectively).	[66]
ZnO	70	*E. coli* 0157:H7	Microbial growth inhibition at ≥12 mmol/L	[67]
ZnO	480–4000	*B. subtilis* CB310, *E. coli* DHSα	Growth inhibition by 90% at 10 mg/L (*B. subtilis*), 48% at 1000 mg/L (*E. coli*)	[59]

(*Continued*)

Table 6.1 Summary of antimicrobial activity of engineered nanoparticles (*cont.*)

ENPs	Size (nm)	Test organism	Effect on microorganism	References
ZnO	<100	*E. coli WP2 trp uvrA*	Induced marginal mutagenesis to organism but no growth inhibition	[38]
ZnO	24–71 and 90–200	*E. coli and V. fischeri*	MIC of 200–250 mg/L and 100 mg/L for *E. coli* and *V. fischeri*, respectively	[68]
ZnO	100–150	*Streptococcus agalactiae, S. aureus*	Inhibition of microbial growth at 0.12 M concentration	[69]
ZnO	50–70	*V. fischeri*	Growth inhibition, bactericidal: 30 min EC_{50} of 1.1–1.9 mg/L at 20 g/L, MIC value of 100 mg/L	[53]
ZnO, ZnO-PVP, ZnO-PS	4,5, 10	*L. monocytogenes, E. coli* 0157:H7, *Salmonella enteritidis*	Dose-dependent bactericidal effect at > 0.3 mg/mL ZnO-PVP, bacteriostatic effect of ZnO powder. No observable inhibitory effect with ZnO-PS	[70]
ZnO and ZnO-Brij-76	10.4 ± 1.2 and 12.7 ± 1.6	*Anabaena flos-aquae* (cyanobacteria)	Size- and capping agent-dependent decrease in photosynthetic activities and cell death. ZnO and ZnO-Brij-76 stimulation of cyanobacterial photosynthetic activity after 10 days	[71]
ZnO	~30	*Campylobacter jejuni, Salmonella enterica serovar enteritidis, E. coli*	MIC of 0.05–0.025 mg/mL (*C. jejuni*), 0.4 mg/mL (*S. enterica* and *E. coli*)	[14]
Ag^0, ZnO, TiO_2	20–40	Activated sludge microbial community	Bactericidal effect on *Nitrosomonas, Nitrobacter, and Nitrospira*	[72]
Ag^0, ZnO, TiO_2	20–40	Methanogens in sewage sludge anaerobic digester	80% reduced abundance and diversity of methanogens with nanotolerant *Methanosarcina acetivorans* and *M. barkeri*; toxic to sulfate-reducing bacteria	[73]

Ag^0, silver oxide nanoparticles; TiO_2, titanium dioxide nanoparticles; ZnO, zinc oxide nanoparticles; SiO_2, silicon dioxide nanoparticles; CeO_2, cerium (iv) oxide nanoparticle; EC_{50}, effective concentration of substance that generates 50% reduction in bioluminescence; β-CD, β-cyclodextrin; TOPO, Tri-*n*-octylphosphine oxide; Brij-76, polyoxyethylene stearyl ether; PVP, polyvinylpyrrolidone; PS, polystyrene; MBC, minimum bactericidal concentration; CO_3O_4, cobalt oxide; Al_2O_2, aluminum oxide; CuO, copper oxide; $TiCl_4$, titanium tetrachloride; Au, gold nanoparticle; LOEC, lowest observed effect concentration; NOEC, no observed effect concentration; MIC, minimum inhibitory concentration.

6.4.3.3 *Effective ENPs toxic dose depends on cultural conditions*

The amount of a substance on a target organism, a system or subpopulation in a specific frequency for a defined duration, commonly called the dose [81], is important in the measurement of toxicity. However, determining the effective dose of ENPs in contact with microbes in a defined or complex medium is a challenge because of potentiation, additive, and synergistic or multiplicative phenomena with micro-components in the medium [82]. ENPs generate different ions in different media, for instance, TiO_2 generated hydroxyl radicals in ultrapure water and superoxide in minimal Davis microbial growth medium (microbial broth used to isolate and characterize nutritional mutants such as *E. coli and B. Subtilis mutants*) [64].

In addition, aerosolized ENPs have been shown to be more toxic than their nanoaggregate form in aquatic media. For example, aerosolized ZnO nanoparticles are more toxic to *S. oneidensis* and *E. coli* two commonly bacterial species used in microbial toxicity than in the ZnO nanoaggregate form [54]. In contrast phosphate buffered solution mitigated ZnO nanoparticles toxic effect on *E. coli* and *S. aureus* growth and viability compared to deionized water and normal saline solution [83]. ZnO nanoparticles were used in these studies because their toxic effect is linked to both photocatalytic potential and ROS production. Furthermore, ENPs effect in batch, continuously fed reactors, aerobic and anaerobic processes [6,72] vary and can influence the effective dose in a particular medium. It is common knowledge that the physicochemical properties and the presence of solids can influence ENPs behavior and reaction [72]. However, aggregated Ag^0 nanoparticles are more toxic to *E. coli* than the dispersed form due to incomplete sulfidation of Ag^0 [84]. The insights from these findings have significantly altered the mode to predict ENPs effect on microbes and raise further questions on how ENPs react and behave in complex medium. Recently, studies have demonstrated that ENPs are selectively toxic to different groups of microorganisms in activated sludge [6,72,85] and on methanogens during anaerobic digestion [73] and this feature was influenced by the complex composition of the environmental media.

6.4.3.4 *Selective toxicity of ENPs on microbial ecophysiology*

The cell wall/membrane maintains structural integrity, osmotic balance and regulates transport essential for the survival of the microorganisms. A disruption of these key functions results in loss of viability and cell death (Fig. 6.1).

The bacterial cell wall composition and their charges [72] can either enhance or attenuate ENP bactericidal effect, although consensus is yet to be established on the relationship between ENPs surface charge and charges on bacterial cell wall. For instance, Ag^0 nanoparticles produce ions (Ag^+) that exert toxic effect on *E. coli* [24] whereas in the presence and absence of fulvic acid with different pH ranges the effect was unrelated with the dissolved Ag^+ [86].

Most studies on the effect of ENPs are focused on pristine nanoparticles and pure cultures of different microbial species in defined medium (Table 6.1). Thus, the empirical

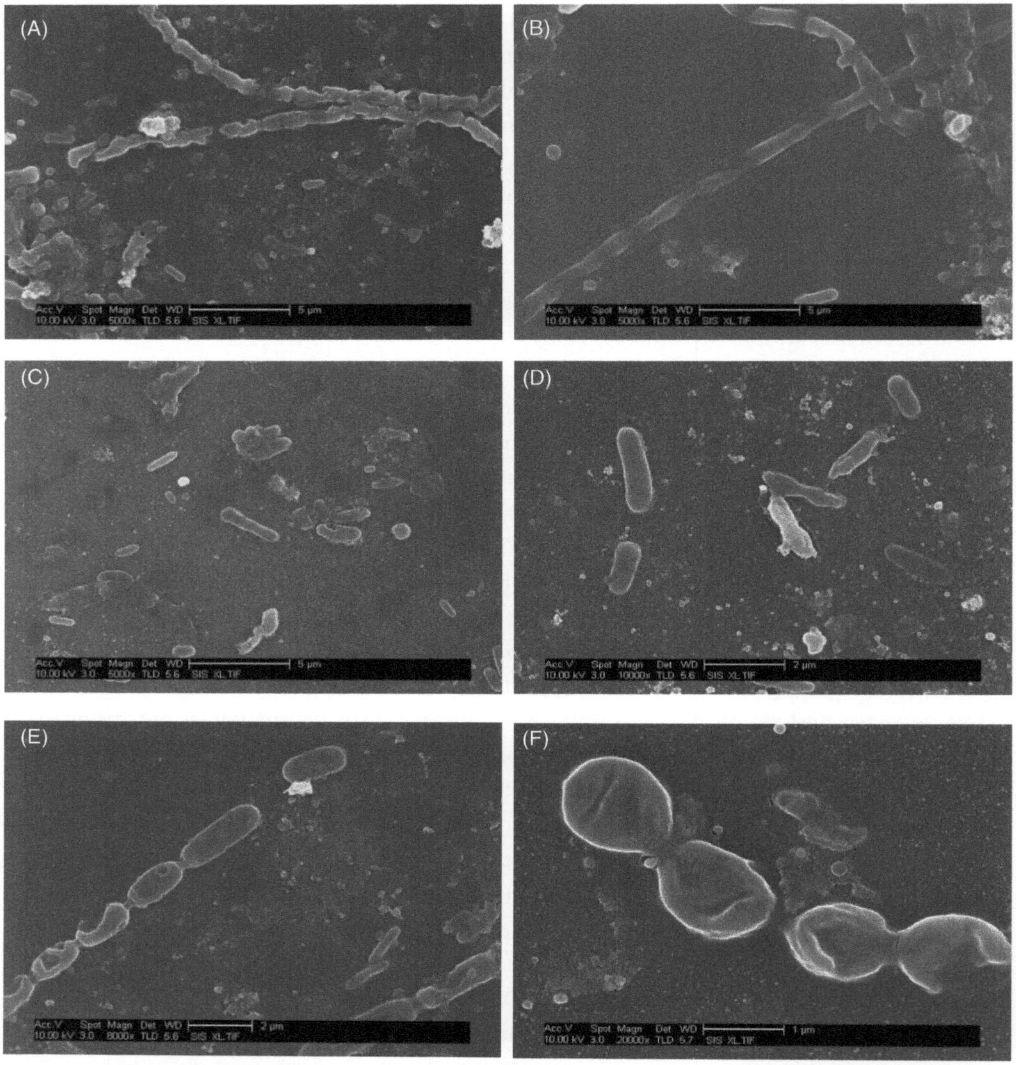

Figure 6.1 ***Effect of ENPs on representative microbial species in activated sludge process.*** **(A,B) Disintegration and dissolution of bacterial cell wall. (C–F) Selective effect of ENPs on different bacterial cells.**

evidence for understanding aged-ENPs effect on microbial community in complex medium is still limited and not clearly defined. Diverse microorganisms found in complex medium respond in different ways to the type and concentration of ENPs [22,58] and are influenced by aggregation which restricts contact and reduces bioavailable dose [87]. For example, fullerenes (C_{60}) were toxic to pure cultures of *B. subtilis* [52] and *E. coli* [88] while no toxic effect was observed on soil bacterial communities due to the presence of NOM [89]. Other studies also reported that different impacts are expected for free-living planktonic and biofilm cells [90]. Typically, there is restricted contact with ENPs in the biofilm by spatial and temporal location of the organisms embedded in extracellular polysaccharides matrix [90]. On the other hand, proteins promote disaggregation of ENP and increase the bioavailable dose [54].

6.4.3.5 Influence of capping agent on ENPs toxic effect

The surface properties of ENPs are one of the most important factors governing their stability, mobility, and/or aggregation into larger particles and deposition in aquatic systems. Stable suspensions of ENPs (e.g., particles rather than aggregates) are in fact a prerequisite for efficient interactions with microorganisms. Great efforts have recently been made toward improving the water stability of ENPs by utilization of surface coatings that alter the solubility, surface charge, and thus their relative hydrophilicity/hydrophobicity [72]. Capping agents protect ENPs from degradation, change their properties (by the addition of catalytically active species and/or specific binding sites), and prevent aggregation due to either charge or steric stabilization mechanisms [18,86]. They may be organic molecules, polymers, biological molecules, or carriers of specific functionalities. For example, metallic ENPs are usually coated with inorganic or organic compounds (e.g., amine, citrate, cysteine, carbonate) or surfactants (e.g., sodium dodecyl sulfate). In some cases, the surface coatings on ENPs have been observed to either enhance or eliminate ENP toxicity to microorganisms. For instance, capping Ag^0 with β-cyclodextrin enhanced the antibacterial properties of silver nanoparticles on *E. coli* ATCC 11229, *Pseudomonas aeruginosa* ATCC 27852, and *S. aureus* ATCC 25923 [18]. Similar findings were found on *Pseudomonas putida* [86] and *Listeria monocytogenes* [70] when ZnO nanoparticles were capped with polyvinylpyrolidone. In contrast, capping ZnO nanoparticles with Brij-76 was initially toxic to cyanobacteria but then stimulated their photosynthetic activity after 10 days exposure [71]. All these bacterial species are indicator organisms commonly used to determine the microbiological quality or safety of water and food as their presence indicates a potential risk of occurrence of pathogenic or toxigenic organisms.

6.4.3.6 Expressing ENPs toxic end points

There are considerably high number of experimental approaches which makes it difficult to determine the effective dose or establish toxicological guidelines for ENPs on microorganisms. In part, this is due to nature of the nanoparticles, inadequate and evolving analytical procedures, exposure time, and physicochemical properties of the sample, and in part, due to the different metabolic end points used to determine toxic effect [4,8]. In addition, ENPs are transformed into by-products that can be toxic or innocuous, or behave and react contrary to Derjaguin–Landau–Verwey–Overbeek (DLVO) theory generally used to interpret and understand how colloidal particles adsorb and desorb from interfaces [4]. Thus, the notionally determined statistical end points such as lethal concentration (LC), lowest observed effect concentration (LOEC), inhibitory concentration (IC), effective concentration (EC), or no observed effect concentration (NOEC) are inadequate to evaluate ENPs toxic effect on microorganisms. These statistical estimates are subjective because hormesis (in toxicology, a concept of biphasic dose response phenomenon by biological systems including microorganisms in which a substance at low dose stimulates and high dose inhibits the activities of the test organism) can occur and the ENPs concentration below the limits of detection exerts biologically significant effect [72]. Overall determining the endpoints of ENPs toxic effect on microorganisms are yet to be further investigated as to date there are inconsistent findings in relation to the varying factors that can influence the toxicology assessment. The effects of ENPs on individual microorganism in pure or mixed culture and on microbial communities vary and depend on several factors such as the type of organisms, the effective dose being applied or the nature of the culture media or environmental matrix being tested [8,87].

6.5 CHALLENGES IN MEASURING AND INTERPRETING ENPs TOXIC EFFECT

It is common knowledge from the various biochemical, physiological, and molecular approaches that most ENPs exhibit inhibitory, toxic, and mutagenic effects on pure cultures of microorganisms. However, it is difficult to extrapolate the effect of ENPs on individual organisms in pure cultures (Table 6.1) to complex microbial communities because the species are diverse in composition, abundance, and activity [8]. In addition, the microbial communities are not uniformly distributed in the environment, the biotic and abiotic characteristics interact in a nonlinear mode and vary with spatial and temporal patterns [90]. Most studies on the effect of ENPs on microorganisms (Table 6.1) indicate in vitro test and the same effect may not be exhibited during in vivo tests or in complex medium such as food, soil, and wastewater with diverse microbial community. In addition, challenges during static and nonstatic renewable toxicity tests; use of continuous and discrete scales of measurement; sampling frame limitations on the measure of diversity; inability to identify or account for confounding factors; and the use of dose–response and time–response approaches are critical challenges.

6.6 CONCLUSIONS

The use of ENPs in consumer products, release into the environmental matrix and effect on nontarget microbes and microbiologically-mediated processes in the ecosystem are of critical concern. Significantly, ENPs and their transformed products differ in reaction from naturally occurring nanoparticles, produce ROS, ion and free radicals that exert toxic effect on microorganisms. The knowledge gap from limited empirical evidence on effect of pristine and aged-ENPs can be bridged by the development of appropriate analytical protocol and framework to evaluate factors that enhance or attenuate ENPs effect. To minimize the uncertainties surrounding how pristine and aged-ENPs toxic effects are interpreted, the toolbox approach involving the use of more than one biological end point to estimate microbial activities, community structure, abundance and diversity with consideration to the confounding factors might be adequate to evaluate microbial response especially in a complex medium. Although recent studies are shedding light on pristine and aged-ENPs toxic effect mechanisms on microorganisms, toxicity data should be evaluated and interpreted on a case-by-case basis for proper understanding.

REFERENCES

[1] C.J. Hurst, Neighborhoods and community involvement: no microbe is an island, in: C.J. Hurst, R.L. Crawford, J.L. Garland, D.A. Lipson, A.L. Mills, L.D. Stetzenbach (Eds.), Manual of Environmental Microbiology, third ed., American Society for Microbiology, Washington, 2007, pp. 6–19.

[2] K.P. Talaro, Foundations in Microbiology, fifth ed., McGraw Hill, NY, (2005) pp. 1–55.

[3] Environmental Protection Agency, EPA, Nanotechnology white paper, US Environmental Protection Agency Report, EPA 100/B-07/001 (2007), Washington.

[4] P. Westerhoff, B. Norwack, Searching for global descriptors of engineered nanomaterial fate and transport in the environment, Acc. Chem. Res. 46 (3) (2013) 844–853.

[5] H. Wigger, S. Hackmann, T. Zimmermann, J. Köser, J. Thöming, A. von Gleich, Influences of use activities and waste management on environmental releases of engineered nanomaterials, Sci. Tot. Environ. 535 (2015) 160–171.

[6] X. Sun, Z. Sheng, Y. Liu, Effect of silver nanoparticles on microbial community structure in activated sludge, Sci. Tot. Environ. 443 (2013) 828–835.

[7] Woodrow Wilson Database. The Project on Emerging Nanotechnologies: Consumer Products Inventory, Woodrow Wilson International Centre for Scholars Washington DC. Available from: http://www.nanoproduct.org/inventories/consumer/

[8] S. Eduok, B. Martins, R. Villa, A. Nocker, B. Jefferson, F. Coulon, Evaluation of the engineered nanoparticle toxic effect on wastewater microorganisms: current status and challenges, Ecotoxicol. Environ. Saf. 9 (2013) 1–9.

[9] Y. Yang, P. Westerhoff, Presence in, and release of, nanomaterials from consumer products, Adv. Exp. Med. Biol. 811 (2014) 1–17.

[10] A.P. Gondikas, A. Morris, B.C. Reinsch, S.M. Marinakos, G.V. Lowry, H. Hsu-Kim, Cysteine-induced modification of zero-valent silver nanomaterials: implications for particle surface chemistry, aggregation, dissolution and speciation, Environ. Sci. Technol. 46 (2012) 7037–7045.

[11] L. Wen-Ru, X. Xiao-Bao, S. Qing-Shan, Z. Hai-Yan, O.-Y. You-Sheng, C. Yi-Ben, Antibacterial activity and mechanism of silver nanoparticles on *Escherichia coli*, Appl. Microbiol. Biotechnol. 85 (2010) 1115–1122.

[12] Y. Ge, J.P. Schimel, P.A. Holden, Evidence for negative effects of TiO_2 and ZnO nanoparticles on soil bacterial communities, Environ. Sci. Technol. 45 (2011) 1659–1664.

[13] K. Tiede, M. Hassellov, E. Breitbarth, Q. Chaudhry, A.B.A. Boxall, Considerations for environmental fate and ecotoxicity testing to support environmental risk assessment for nanoparticles, J. Chromatogr. A 1216 (2009) 503–509.

[14] Y. Xie, Y. He, P.L. Irwin, T. Jin, X. Shi, Antibacterial activity and mechanism of action of zinc oxide nanoparticles against *Campylobacter jejuni*, Appl. Environ. Microbiol. 77 (7) (2011) 2325–2331.

[15] I. Rodea-Palomares, K. Boltes, F. Fernandez-Pinas, F. Leganes, E. Garcia-Calvo, J. Santiago, R. Rosal, Physicochemical characterization and ecotoxicological assessment of CeO_2 nanoparticle using two aquatic microorganisms, Toxicol. Sci. 119 (1) (2011) 135–145.

[16] J. Wei, H. Mashayekhi, B. Xing, Bacterial toxicity comparison between nano- and micro-scaled oxide particles, Environ. Pollut. 157 (2009) 1619–1625.

[17] A.K. Suresh, D.A. Pelletier, W. Wang, J. Moon, B. Gu, N.P. Mortensen, D.P. Allison, D.C. Joy, T.J. Phelps, M.J. Doktycz, Silver nanocrystallites: biofabrication using *Shewanella oneidensis*, and an evaluation of their comparative toxicity on gram-negative and gram-positive bacteria, Environ. Sci. Technol. 44 (13) (2010) 5210–5215.

[18] S. Jaiswal, B. Duffy, A.K. Jaiswal, N. Stobie, P. McHale, Enhancement of the antibacterial properties of silver nanoparticles using β-cyclodextrin as a capping agent, Int.J.Antimicrob.Agent 36 (2010) 280–283.
[19] A. Nabikhan, K. Kandasamy, A. Raj, N.M. Alikunhi, Synthesis of antimicrobial silver nanoparticles by callus and leaf extracts from saltmarsh plant *Sesuvium portulacastrum* L., Colloids Surf. B 79 (2) (2010) 488–493.
[20] L.F. Espinosa-Cristobal, G.A. Martinez-Castanon, R.E. Martinez-Martinez, J.P. Loyola-Rodriguez, N. Patino-Martin, J.F. Reyes-Macias, F. Ruiz, Antibacterial effect of silver nanoparticles against *Streptococcus mutans*, Mater. Lett. 63 (2009) 2603–2606.
[21] C. Lok, C. Ho, R. Chen, Q. He, W. Yu, H. Sun, P.K. Tam, J. Chiu, C. Che, Silver nanoparticles: partial oxidation and antibacterial activities, J. Biol. Inorg. Chem. 12 (4) (2007) 527–534.
[22] O. Choi, K.K. Deng, N.-J. Kim, L. Ross Jr., R.Y. Surampalli, Z. Hu, The inhibitory effect of silver nanoparticles, silver ions, and silver chloride colloids on microbial growth, Wat. Res. 42 (12) (2008) 3066–3074.
[23] N.V. Ayala-Nunez, H.H. Lara, L.C.I. Turrent, C.R. Padilla, Silver nanoparticles toxicity and bactericidal effect against methicillin-resistant *Staphylococcus aureus*: nanoscale does matter, Nanobiotechnology 5 (2009) 2–9.
[24] I. Sondi, B. Salopek-Sondi, Silver nanoparticles as antimicrobial agent: a case study on *E. coli* as a model for Gram-negative bacteria, J. Colloid Interface Sci. 275 (2004) 177–182.
[25] F. Martinez-Gutierrez, P.L. Olive, A. Banuelos, E. Orrantia, N. Nino, E.M. Sanchez, F. Ruiz, H. Bach, Y. Av-Gay, Synthesis, characterization, and evaluation of antimicrobial and cytotoxic effect of silver and titanium nanoparticles, Nanomedicine 6 (2010) 681–688.
[26] V.C. Verma, R.N. Kharwar, A.C. Gange, Biosynthesis of antimicrobial silver nanoparticles by the endophytic fungus *Aspergillus clavatus*, Nanomedicine 5 (1) (2010) 33–40.
[27] S. Sadhasivam, P. Shanmugam, K. Yun, Biosynthesis of silver nanoparticles by *Streptomyces hygroscopicus* and antimicrobial activity against medically important pathogenic microorganisms, Colloids Surf. B 81 (2010) 325–362.
[28] C. Krishnaraj, E.G. Jagan, S. Rajasekar, P. Selvakumar, P.T. Kalaichelvan, N. Mohan, Synthesis of silver nanoparticles using *Acalypha indica* leaf extracts and its antibacterial activity against water borne pathogens, Colloids Surf. B 76 (1) (2010) 50–56.
[29] N. Drogat, R. Granet, V. Sol, A. Memmi, N. Saad, C.K. Koerkamp, P. Bressollier, P. Krausz, Antimicrobial silver nanoparticles generated on cellulose nanocrystals, J. Nanopart. Res. 3 (4) (2010) 1557–1562.
[30] J. Thiel, L. Pakstis, S. Buzby, M. Raffi, C. Ni, D.J. Pochan, S.I. Shah, Antimicrobial properties of silver-doped titania, Small 3 (2007) 799–803.
[31] G.A. Martinez-Castanon, N. Nino-Martinez, F. Martinez-Gutierrez, J.R. Martinez-Mendoza, F. Ruiz, Synthesis and antibacterial activity of silver nanoparticle with different sizes, J. Nanopart. Res. 10 (2008) 1343–1348.
[32] G.K. Vertelov, Y.A. Krutyakov, A.Y. Efremenkova, O.V. Olenin, G.V. Lisichkin, A versatile synthesis of highly bactericidal Myramistin® stabilized silver nanoparticles, Nanotechnology 19 (35) (2008) 355707.
[33] B.C. Lee, S. Kim, H.K. Shon, S. Vigneswaran, S.D. Kim, J. Cho, I.S. Kim, K.H. Choi, J.B. Kim, H.J. Park, J. Kim, Aquatic toxicity evaluation of TiO_2 nanoparticle produced from sludge of $TiCl_4$ flocculation of wastewater and seawater, J. Nanopart. Res. 11 (8) (2009) 2087–2096.
[34] A. Bradford, R.D. Handy, J.W. Readman, A. Atfield, M. Mühling, Impact of silver nanoparticle contamination on the genetic diversity of natural bacterial assemblages in estuarine sediments, Environ. Sci. Technol. 43 (12) (2009) 4530–4536.
[35] H. Zhang, G. Chen, Potent antibacterial activities of Ag/TiO_2 nanocomposite powders synthesized by a one-pot sol-gel method, Environ. Sci. Technol. 43 (2009) 2905–2910.
[36] A. Melaiye, Z. Sun, K. Hindi, A. Milsted, D. Ely, D.H. Reneker, C.A. Tessier, W.J. Youngs, Silver (I) imidazole cyclophane gem-diol complexes encapsulated by electrospun tecophilic nanofibres: formation of nanosilver particles and antimicrobial activity, J. Am. Chem. Soc. 127 (2005) 2285–2291.
[37] H.H. Lara, N.V. Ayala-Nunez, L.C.I. Turrent, C.R. Padilla, Bacterial effect of silver of nanoparticles against multidrug-resistant bacteria, World J. Microbiol. Biotechnol. 26 (2010) 615–621.
[38] X. Pan, J.E. Redding, P.A. Wiley, L. Wen, J.S. McConnell, B. Zhang, Mutagenicity evaluation of metal oxide nanoparticles by bacterial reverse mutation assay, Chemosphere 79 (2010) 113–116.

[39] M.-H. Park, K.-H. Kim, H.H. Lee, J.-S. Kim, S.-J. Hwang, Selective inhibitory potential of silver nanoparticles on the harmful cyanobacterium *Microcystis aeruginosa*, Biotechnol. Lett. 32 (2010) 423–428.
[40] B. Chudasama, A.K. Vala, N. Andhariya, R.V. Mehta, R.V. Upadhyay, Highly bacterial resistant silver: synthesis and antibacterial activities, J. Nanopart. Res. 12 (2010) 1677–1685.
[41] K. Kalishwaralal, S. BarathManiKanth, S.R. Pandian, K. Deepak, S. Gurunathan, Silver nanoparticles impede the biofilm formation by *Pseudomonas aeruginosa* and *Staphylococcus epidermidis*, Colloids Surf. B 79 (2) (2010) 340–344.
[42] L.R. Jaidev, G. Narasimha, Fungal mediated biosynthesis of silver nanoparticles, characterization and antimicrobial activity, Colloids Surf. B 81 (2010) 430–433.
[43] A. Bankar, B. Joshi, R.K. Kumar, S. Zinjarde, Banana peel extract mediated novel route for the synthesis of silver nanoparticles, Colloids Surf. A 368 (2010) 58–63.
[44] P. Gajjar, B. Pettee, D.W. Britt, W. Huang, W.P. Johnson, A.J. Anderson, Antimicrobial activities of commercial activities nanoparticles against an environmental soil microbe, *Pseudomonas putida* KT2440, J. Biol. Eng. 3 (2009) 1754-1611.
[45] V. Gopinatha, D. MubarakAlib, S. Priyadarshinia, N. Meera Priyadharsshinia, N. Thajuddinb, P. Velusamya, Biosynthesis of silver nanoparticles from *Tribulus terrestris* and its antimicrobial activity: a novel biological approach, Colloids Surf. B 96 (2012) 69–74.
[46] S.W. Kim, Y.-W. Baek, Y.-J. An, Assay-dependent effect of silver nanoparticle to *Escherichia coli* and *Bacillus subtilis*, Appl. Microbiol. Biotechnol. 92 (2011) 1045–1052.
[47] M.R. Das, R.K. Sarma, R. Saikia, V.S. Kale, M.V. Shelke, P. Sengupta, Synthesis of silver nanoparticles in an aqueous suspension of graphene oxide sheets and its antimicrobial activity, Colloids Surf. B 83 (1) (2011) 16–22.
[48] R. Barrena, E. Casals, J. Colon, X. Font, A. Sanchez, V. Puntes, Evaluation of the ecotoxicity of model nanoparticles, Chemosphere 75 (2009) 850–857.
[49] I.M. Sadiq, B. Chowdhury, N. Chandrasekaran, A. Mukherjee, Antimicrobial sensitivity of *Escherichia coli* to alumina nanoparticles, Nanomedicine 5 (3) (2009) 282–286.
[50] N. Strigul, L. Vaccari, C. Galdun, M. Wazne, X. Liu, C. Christodoulatos, K. Jasinkiewicz, Acute toxicity of boron, titanium dioxide and aluminium nanoparticles to *Daphnia magna* and *Vibrio fischeri*, Desalination 248 (1-3) (2009) 771–782.
[51] A. Garcia, R. Espinosa, L. Delgado, E. Casals, E. Gonzalez, V. Puntes, C. Barata, X. Font, A. Sanchez, Acute toxicity of cerium oxide, titanium oxide and iron nanoparticles using standardized tests, Desalination 269 (1-3) (2011) 136–141.
[52] D.Y. Lyon, J.D. Fortner, C.M. Sayes, V.L. Colvin, J.B. Hughes, Bacterial cell association antimicrobial activity of a C-60 water suspension, Environ. Toxicol. Chem. 24 (11) (2005) 2757–2762.
[53] M. Heinlaan, A. Ivask, I. Blinova, H. Dubourguier, A. Kahru, Toxicity of nanosized and bulk ZnO, CuO and TiO_2 to bacteria *Vibrio fischeri* and crustaceans *Daphnia magna* and *Thamnocephalus platyurus*, Chemosphere 71 (7) (2008) 1308–1316.
[54] B. Wu, Y. Wang, Y.-H. Lee, A. Horst, Z. Wang, D.-R. Chen, R. Sureshkumar, Y.J. Tang, Comparative ecotoxicities of nano-ZnO particles under aquatic and aerosolized exposure modes, Environ. Sci. Technol. 44 (2010) 1484–1489.
[55] S. Chatterjee, A. Bandyopadhyay, K. Sarkar, Effect of iron oxide and gold nanoparticles on bacterial growth leading towards biological application, J. Nanobiotechnol. 9 (2011) 34.
[56] V. Shah, P. Dobiasova, P. Baldrian, F. Nerud, A. Kumar, S. Seal, Influence of iron and copper nanoparticle powder on the production of lignocellulose degrading enzymes in the fungus Trametes versicolor, J. Hazard Mater. 178 (2010) 1141–1145.
[57] H.D. Jang, S.-K. Kim, S.-J. Kim, Effect of particle size and phase composition of titanium dioxide nanoparticles on the photocatalytic properties, J. Nanopart. Res. 3 (2007) 141–147.
[58] L.K. Adams, D.Y. Lyon, P.J.J. Alvarez, Comparative eco-toxicity of nanoscale TiO_2, SiO_2, and ZnO water suspensions, Wat. Res. 40 (19) (2006) 3527–3532.
[59] M. Schaefer, J. Scott-Fordsmand, Nanoparticles: does size matters? in: 16th Annual Meeting of SETAC Europe, The Hague, Netherlands, 7–11 May 2006, p. 222.
[60] G. Fu, P.S. Vary, C.T. Lin, Anatase TiO_2 nanocomposites for antimicrobial coating, J. Phys. Chem. B 109 (2005) 8889–8898.

[61] A. Kumar, A.K. Pandey, S.S. Singh, R. Shanker, A. Dhawan, Cellular uptake and mutagenic potential of metal oxide nanoparticles in bacterial cells, Chemosphere 83 (8) (2011) 1124–1132.

[62] A. Ivask, O. Bondarenko, N. Jepihhina, A. Kahru, Profiling of the reactice oxygen species-related ecotoxicity of CuO, ZnO, TiO_2, silver and fullerene nanoparticles using a set of recombinant luminescent *Escherichia coli* strains: differentiating the impact of particles and solubilised metals, Anal. Bioanal. Chem. 398 (2010) 701–716.

[63] K. Kasemets, A. Ivask, H.-C. Dubourguier, A. Kahru, Toxicity of nanoparticles of ZnO, CuO and TiO_2 to yeast *Saccharomyces cerevisiae*, Toxicol. In Vitro 23 (2009) 1116–1122.

[64] L. Brunet, D.Y. Lyon, E.M. Hotze, P.J.J. Alvarez, M.R. Wiesner, Comparative photoactivity and antibacterial properties of C60 fullerenes and titanium dioxide nanoparticles, Environ. Sci. Technol. 43 (12) (2009) 4355–4360.

[65] M. Diao, M. Yao, Use of zero-valent iron nanoparticles in inactivating microbes, Wat. Res. 43 (2009) 5243–5251.

[66] M. Banoee, S. Seif, Z.E. Nazari, P. Jafari-Fesharaki, H.R. Shahverdi, A. Moballegh, K.M. Moghaddam, A.R. Shahverdi, ZnO nanoparticles enhanced antibacterial activity of ciprofloxacin against *Staphyloccocus aureus* and *Escherichia coli*, J. Biomed. Mater. Res. B 93 (2) (2010) 557–561.

[67] Y. Liu, L. He, A. Mustapha, H. Li, Z.Q. Hu, M. Lin, Antibacterial activities of zinc nanoparticles against *Escherichia coli* O157:H7, Appl. Microbiol. 107 (4) (2009) 1193–1201.

[68] L. Zhang, Y. Jiang, Y. Ding, M. Povey, D. York, Investigation into the antibacterial behaviour of suspensions of ZnO nanoparticles (ZnO nanofluids), J. Nanopart. Res. 9 (2007) 479–489.

[69] Z. Huang, X. Zheng, D. Yan, G. Yin, X. Liao, Y. Kang, Y. Yao, D. Huang, B. Hao, Toxicological effects of ZnO nanoparticles based on bacteria, Langmuir 24 (2008) 4140–4144.

[70] T. Jin, D. Sun, J.Y. Su, H. Zhang, H.J. Sue, Antimicrobial efficacy of zinc oxide quantum dots against *Listeria monocytogenes, Salmonella enteritidis* and *Escherichia coli* 0157:H7, J. Food Sci. 74 (1) (2009) 46–52.

[71] R. Brayner, S.A. Dahoumane, C. Yéprémian, C. Djediat, M. Meyer, A. Couté, F. Fiévet, ZnO nanoparticles: synthesis, characterization, and ecotoxicological studies, Langmuir 26 (9) (2010) 6522–6528.

[72] S. Eduok, C. Hendry, R. Ferguson, B. Martin, R. Villa, B. Jefferson, F. Coulon, Insights into the effect of mixed engineered nanoparticles on activated sludge performance, FEMS Microbiol. Ecol. 91 (2015) 1–9.

[73] S. Eduok, Evaluation of the impact of engineered nanoparticles on the operation of wastewater treatment plant, PhD Thesis, Cranfield University, UK, 2013.

[74] O. Choi, C.-P. Yu, E. Fernandez, Z. Hu, Interaction of nanosilver with *Escherichia coli* cells in planktonic biofilms cultures, Wat. Res. 44 (2010) 6095–6103.

[75] S. Pal, Y.K. Tak, J.M. Song, Does the antibacterial activity of silver nanoparticles depend on the shape of the nanoparticles? A study of the gram-negative bacterium *Escherichia coli*, Appl. Environ. Microbiol. 73 (6) (2007) 1712–1720.

[76] A.P. Gondikas, A. Morris, B.C. Reinsch, S.M. Marinakos, G.V. Lowry, H. Hsu-Kim, Cysteine-induced modification of zero-valent silver nanomaterials: implications for particle surface chemistry, aggregation, dissolution, and speciation, Environ. Sci. Technol. 46 (2012) 7037–7045.

[77] N. Kucerka, E. Papp-Szabo, M.P. Njeh, T.A. Harroun, S.R. Schooling, J. Pencer, E.A. Nicholson, T.J. Beveridge, J. Katsaras, Effects of cation on the structure of bilayers formed by lipopolysaccharides from *Pseudomonas aeruginosa* PAOI, J. Phys. Chem. B 112 (27) (2008) 8057–8062.

[78] C. Beer, R. Foldbjerg, Y. Hayashi, D.S. Sutherland, H. Autrup, Toxicity of silver nanoparticles—nanoparticle or silver ions?, Toxicol. Lett. 202 (2012) 286–292.

[79] E. Cabiscol, J. Tamarit, J. Ros, Oxidative stress in bacteria and protein damage by reactive oxygen species, Int. Microbiol. 3 (2000) 3–8.

[80] J.L. Elechiguerra, J.L. Burt, J.R. Morones, A. Camacho-Bragado, X. Gao, H.H. Lara, M.J. Yacaman, Interaction of silver nanoparticles with HIV-1, J. Nanobiotechnol. 3 (2005) 6.

[81] C.V.J. Leeuwen, T. Vermeire, T.G. Vermeire, Risk Assessment of Chemicals: An Introduction, second ed., Springer-Verlag, Netherland, (2007).

[82] A.M. El-Badawy, T.P. Luxton, R.G. Silva, K.G. Scheckel, M.T. Suidan, T.M. Tolaymat, Impact of environmental conditions (pH, ionic strength, and electrolyte type) on the surface charge and aggregation of silver nanoparticles suspensions, Environ. Sci. Technol. 44 (4) (2010) 1260–1266.

[83] S. Eduok, N. Asamudo, Diluents mitigates the inhibitory effect of zinc oxide nanoparticles on *Escherichia coli* and *Staphylococcus aureus*, Br. Microbiol. Res. J. 12 (5) (2016) 1–8.
[84] B.C. Reinsch, C. Levard, Z. Li, R. Ma, A. Wise, K.B. Gregory, G.E. Brown Jr., G.V. Lowry, Sulfidation decreases silver nanoparticle growth inhibition effects for *Escherichia coli*, Environ. Sci. Technol. 46 (2012) 6992–7000.
[85] A. Garcia, L. Delgado, J.A. Tora, E. Casals, E. Gonzalez, V. Puntes, X. Font, J. Carrera, A. Sanchez, Effect of cerium, titanium dioxide, silver, and gold nanoparticles on the activity of microbial communities intended in wastewater treatment, J. Hazard Mater. 199-200 (2012) 64–72.
[86] J. Fabrega, S.R. Fawcett, J.C. Renshaw, J.R. Lead, Silver nanoparticle impact on bacterial growth: effect of pH, concentration, and organic matter, Environ. Sci. Technol. 43 (19) (2009) 7285–7290.
[87] J. Fabrega, S.N. Luoma, C.R. Tyler, T.S. Galloway, J.R. Lead, Silver nanoparticles: behaviour and effects in the aquatic environment, Environ. Inter. 37 (2) (2011) 517–531.
[88] S.R. Chae, S. Wang, Z. Hendren, M.R. Weisner, Y. Watanabe, C.K. Gunsch, Effects of fullerene nanoparticles on *Escherichia coli* k12 respiratory activity in aqueous suspension and potential use for membrane biofouling control, J. Membr. Sci. 329 (1-2) (2009) 68–74.
[89] Z. Tong, M. Bischoff, L. Niles, B. Applegate, R.F. Turco, Impact of fullerene (C_{60}) on a soil microbial community, Environ. Sci. Technol. 41 (2007) 2985–2991.
[90] M.A. Kiser, H. Ryu, H. Jang, K. Hristovski, P. Westerhoff, Biosorption of nanoparticles to heterotrophic wastewater biomass, Wat. Res. 44 (2010) 4105–4114.

CHAPTER SEVEN

Polymer Nanocomposites for Food Packaging

Joseba Luna*, Alejandro Vílchez**
*ONYRIQ, Barcelona, Spain
**LEITAT Technological Centre, Terrassa, Spain

Contents

7.1 INTRODUCTION

Polymers have replaced conventional materials in packaging applications due to their functionality, lightweight, ease of processing, and low cost. The use of synthetic polymers is ubiquitous in food packaging where they provide mechanical, chemical, and microbial protection from the environment and allow product display. The chemical composition of polymers can be tailored to achieve a variety of properties; therefore plastics are versatile materials. However, they present several limitations regarding the gas barrier properties, especially, since it is difficult to achieve a good balance of gas and water vapor transmission. Depending on the application, such as fresh food containment, the lack of gas barrier behavior can compromise the shelf life of the contained food, as well as its organoleptic properties. This drawback, among other performance limitations, such as lack of UV protection or limited printability, is usually overcome by producing multilayered systems in which the different layers have specific functionality.

Emerging Nanotechnologies in Food Science

Multilayer systems present outstanding performance for desired specifications, such as mechanical resistance, printability, or barrier performance. The complexity of multilayer structures greatly hinders their recyclability in cases where the layers are bonded with adhesives that cannot be removed or are bonded by direct melt joint produced by coextrusion processes. In addition, the need for several layers directly affects the material costs in terms of weight and material consumption and, hence, the associated carbon footprint of the entire value chain. Therefore, the applied material science research community is looking for new implementations of technologies that can overcome the barriers of the current in-market products. Within this framework, nanotechnology meets packaging. According to market estimates, food packaging applications make up the largest share of the current and short-term predicted market for nanoenabled products in the food sector [1].

Nanotechnology has brought improvement to polymeric materials since the mid 1980s [2]. However, the true start of the nanocomposites' history can be dated back to 1990, when Toyota first used a clay/nylon-6 nanocomposite to produce timing belt covers [3], replacing the metal parts previously used to protect this section of the car engine. The early-noticed improvement of the barrier properties [4,5], as well as mechanical behavior [6], along with the low dosage of nanomaterials needed to achieve this enhancement in the material properties, generated interest in the use of polymer nanocomposites in packaging applications. Nanotechnology in food packaging enables the altering of the structure of the packaging materials on the molecular scale to give the materials desired properties. Simple traditional "packing" is to be replaced with multifunctional intelligent packaging methods, which can provide advantages such as extended shelf life, sensoring the quality of the product, or bactericidal properties. New packaging structures that include nanotechnology have new functionalities that can improve food quality. They can also be lightweight solutions with reduced cost in terms of production and transportation. Polymer nanocomposites consist of a polymeric continuous matrix containing one or more doping agents (commonly called fillers) whose individual elements have at least one of their dimensions in the nanoscale range [7]. Nanocomposites have emerged as novel food packaging materials due to their benefits such as enhanced mechanical, thermal, and barrier properties.

This chapter aims to depict an overview of the different polymer nanocomposites suitable for food packaging applications in terms of production, properties enhancement, risks, and future trends.

7.2 PREPARATION AND TRANSFORMATION OF POLYMER NANOCOMPOSITES FOR FOOD PACKAGING

7.2.1 The chemistry behind polymer nanocomposites

7.2.1.1 The role of the interface

The interface between the polymer and nanoparticles is the key factor of nanocomposite final properties. According to Kickelbick, a cube of 16 × 16 × 16 packed atoms

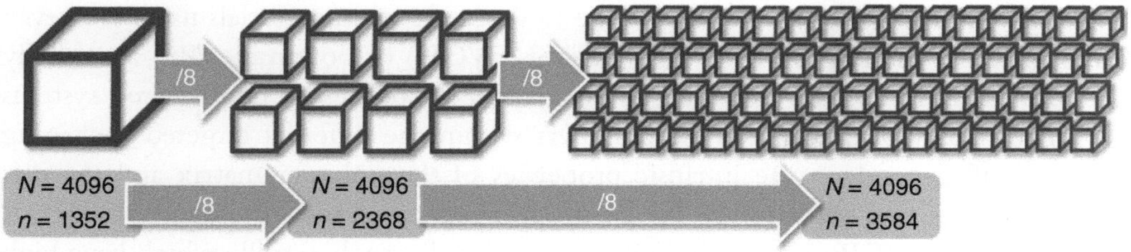

Figure 7.1 ***Effect of the subdivision of a cubic particle on surface area.*** **"*N*" refers to the total number of atoms, and "*n*" indicates the number of atoms at the surface of the particles.**

contains an overall number of 4096 atoms. Thus, 1352 are located on surface comprising around 33% of the total [8]. If the cube is divided into eight equal parts, the overall atom number remains the same; however the number of atoms at the surface increases to 2368 representing 58% of the total. By increasing the cube's subdivision we reach 3584 surface atoms, which supposes an 88% of surface atoms (Fig. 7.1). The newly exposed inner interface will have a direct impact on material properties. Based on the interfacial chemistry two distinct categories of interfaces can be defined; the first one consists of weak bonds such as hydrogen bonding, Van der Waals, or ionic interaction, whereas the second is based on covalent bonds between organic and inorganic components [9–11].

Several research works have been published on the preparation of nanocomposites since they were first discovered; however, achieving homogeneous dispersion of the nanoparticles in polymeric matrices still remains a difficulty [12]. In most cases, interactions based on small forces such as Van der Waals attraction between nanoparticles promote formation of agglomerates. In addition, hydrophilic nanoparticles and hydrophobic polymers are not fully stable together, which results in poor interfacial adhesion and consequently incomplete dispersion and limited properties. Hence, nanocomposites can exhibit worse properties than conventional polymers without nanomaterials; this limits their effective application [13,14]. To overcome these interfacial problems, scientists have focused on synthesis processes, as well as development of new combined methods to achieve higher control on the composition of the polymers and resulting morphology. Additionally, functionalization of particles or innovative procedures, such as reactive extrusion which consists of chemical reactions triggered by high shear stress and temperature, and promoted by an initiator toward creating covalent bonds between the continuous matrix and dispersed domains [15], are currently under development to achieve all the benefits that nanotechnology can bring to polymer composites; these mentioned pathways will be further discussed in this chapter.

7.2.1.2 Main polymer matrices for polymer nanocomposites

Polymer food packaging represents 30% of the total share of the polymer industry [16], comprising almost a fifth of its net revenue. High-density polyethylene (HDPE), low-density polyethylene (LDPE), polypropylene (PP), polyethylene terephthalate (PET), polyvinyl chloride (PVC), and polystyrene (PS) are the main materials used. However, other materials such as ethylene vinyl alcohol (EVOH) or polyamide (PA) are usually employed mainly because they can confer barrier properties to multilayered systems [17]. The final application of these polymers within the different targeted packaging product will depend on the intrinsic properties of the polymer matrix and the different additives that can modify the final performance of the developed product. A polymer such as HDPE is used for containers or bottles, such as milk, which have high demand, in terms of moisture barrier, chemical resistance, and strong mechanical properties. However, transparency is not among its requirements [18]. Meanwhile, LDPE is used for food storage bags because it is cost effective and has a large stretch capacity as well as excellent barrier properties [19]. Another polymer belonging to the family of polyolefins is PP, which is used in rigid containers when high strength properties are needed, though it is slightly more expensive than the aforementioned polyethylene family [20]. PET presents very good tensile and yield strength properties, as well as clarity and high oxygen barrier properties after processing. Therefore, it is broadly used for cold beverages [21]. PVC is most commonly used for clear plastic wrapping because of its well-known cost effectiveness and great stretching capabilities [22]. However, due to the plasticizers needed for its appropriate processing behavior [23], usually esters of carboxylic acids such as phthalates or adipates, PVC is likely to disappear within the different primary packaging structures in the next few years, unless migration problems from the plasticizers to the food are solved [24,25]. Finally, PS is commonly used in its foamed format for food containers, as well as meat and egg trays that require a rigid form [26]. Combinations of these different materials in multilayered systems can lead to packaging solutions with enhanced properties that can be used for a variety of applications, such as water jugs or oven-bake bags.

These conventional polymers are also the ones whose properties have been modified the most by means of nanoadditivation [27]. In this way, the vast majority of the studies have been mainly focused on the enhancement of polyolefin's properties since the 1990s [28], principally, with the aim of improving oxygen barrier, mechanical performance, and thermal resistance of both HDPE [29] and PP [30]. Concerning PET, the greater part of the publications tackle barrier performance as the target property to be improved [31]. Additionally, other materials and their nanocomposites, such as polyamide, are emerging as feasible solutions to bring enhanced properties to food packaging products [32].

It is also important to highlight that there is not any current packaging plastic that addresses other problems, such as biodegradability or petrol resource consumption in its preparation. Within this framework, the environmental impact of persistent plastic packaging waste is raising global concern as disposal methods are limited. Furthermore, exhausting natural scarce resources by manufacturing fossil fuel-based plastic packaging materials has increased the demand for biodegradable polymers from renewable sources as an alternative to their petrol-based counterparts, especially when it comes to short-term and disposable packaging applications [33,34].

Bio-based polymers are those in which constitutional units are wholly or partly from biomass origin [35]. The development of these products entails many different benefits; like preservation of resources, reduction of the carbon footprint and the petroleum dependence with new alternatives, enhancement of farm and rural economies, and creation of green jobs, as documented by Wolf et al. [36]. Meanwhile, biodegradable polymers are defined as materials in which at least one step in the degradation process takes place through metabolism of naturally occurring organisms [37]. Under appropriate conditions of moisture, temperature, and oxygen concentration, on paper, biodegradation leads to no environmentally hazardous residue or by-products [38]. Bio-based polymers can be classified into different groups based on their manufacturing processes. They include (1) synthetic biodegradable polymers such as poly(L-lactide) (PLA), poly(ε-caprolactone) (PCL), poly(glycolic acid) (PGA), poly(butylenes succinate) (PBS), etc. (2) biopolymers produced by microbial fermentation like microbial polyesters, such as poly(hydroxyalkanoates) (PHAs) including poly(β-hydroxybutyrate) (PHB), poly(3-hydroxyvalerate) (PHV), poly(3-hydroxyhexanoate) (PHH), and copolymers thereof; and (3) natural biopolymers such as starch, cellulose, chitosan, etc. [39–41]. Among these materials, PLA, PHAs, and starch are the prevailing choices for packaging applications. This is because they can be used as raw materials or as the matrix of a polymer nanocomposite with enhanced properties to overcome the inherent shortcomings of these materials such as limited mechanical performance, difficulty of processability, poor thermal resistance, or hydrolysis sensitivity [42–44].

7.2.1.3 Nanoparticles for polymer nanocomposites

Nanoplatelets composed of clays or other silicates materials are the most promising nanoscale fillers for nanocomposites. The popularity of nanoclay in food contact applications derives from their low cost, low dosage needed, high stability, and (alleged) safety [45]. The first studies regarding clay/polymer nanomaterials for packaging usage date back to the last decade of the 20th century, when several patents were registered regarding the dispersion of layered silicates within different polymeric matrices. Different production routes were implemented like in situ polymerization or dispersion processes [46,47]. The use of organo-modified clays has been a broadly studied area [48,49]: their lamellar structure (Fig. 7.2) can confer singular behavior to the final composite, which is very appropriate for their application as nanofiller [50]. Additionally, clays provide versatility [51,52] for further chemical modification, which enables the compatibility with different polymeric domains [53] by having direct impact on the polymer–clay interfacial tension and improving the dispersion of filler in the polymer matrix [54].

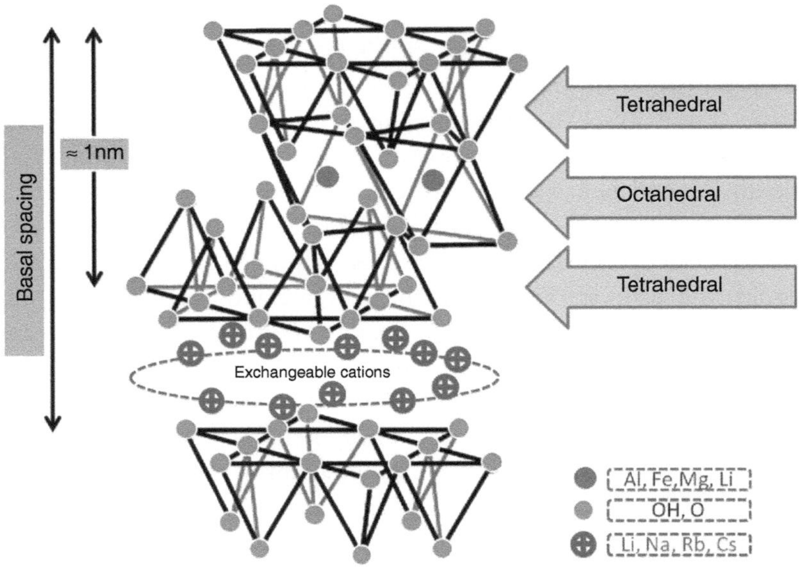

Figure 7.2 ***Structure of layered silicates.***

Improvement of barrier properties, as well as mechanical or fire resistance, are some of the developments that have been reported in several studies with different polymeric matrices [55,56].

Another nanofiller which has gained interest during the last few years is nanocellulose in its different formats. Two types of nanoreinforcements can be obtained starting from cellulose [57]. In nature, cellulose chains form microfibrils (or nanofibers), bundles of polysaccharide chains held together by hydrogen bonding [58]. Each microfibril consists of aggregated elementary nanofibrils, which present crystalline and amorphous domains. The crystalline parts, known as nanowhiskers, nanorods, or nanocrystals, can be isolated by different treatments. In brief, these treatments involve removing the amorphous domains present in the fibrils, and thus exposing the crystalline regions. This process does not damage the crystalline regions because they present higher chemical resistance than the amorphous regions [59]. The lengths thereof can vary from 500 nm up to 1–2 μm, and their diameter can go down to about 8–20 nm, namely resulting in high aspect ratios, which is fundamental to their proper performance [60,61]. Back in 1996, Helbert et al. [62] reported having prepared crystalline nanofibrils whose elastic modulus was only about seven times lower than those of single-walled carbon nanotubes (CNTs) reported by Podsiadlo et al. [63]. On the other hand, no conclusive results were achieved with respect to elongation at break. The results reported by Dogan et al. [64] suggest that the cellulosic reinforcement hinders the elongation observed in the matrix without reinforcement. Still other works uphold that elongation increases [65], or at least no significant changes can be expected [66], from the addition of cellulose nanofibers. Moisture barrier of polymer has also been reported to be improved by Paralikar et al. [67], due to slower diffusion processes [68].

Carbon-based nanofillers, such as CNTs or graphene, have come to the fore since the start of the 21st century [69,70]. CNTs consist of a one-atom thick single-wall nanotube (SWNT), or a number of concentric tubes known as multiwalled nanotubes (MWNT), having extraordinary high aspect ratios and elastic modulus [71]. Their principal characteristic is their high Young's modulus and tensile strength values that can reach as high as 1 TPa and 200 GPa, respectively [72]. In addition, the fact that their surface can be derivatized makes them perfect candidates to be dispersed within a polymer matrix [73]. Even in minimum concentrations, CNTs have proven to greatly enhance thermal stability as well as tensile strength and elastic modulus of polymers, such as PVOH [74], PP, or PA [75]. Similar to CNTs, graphene has emerged as a feasible nanofiller to improve polymer performance in terms of mechanical resistance or barrier properties due to its layered structure [70] and ease of their chemical modification [76]. Thus, recent studies carried out by Xiang et al. reported that gas effective diffusivity of a polyurethane matrix decreased by three orders of magnitude with only 0.5% in weight of graphene [77].

Silica nanoparticles ($nSiO_2$) have been demonstrated to improve mechanical and barrier properties of polymer matrices. The addition of $nSiO_2$ into PP improved the tensile properties of the material in terms of strength, modulus, and also elongation [78]. This phenomenon has been reproduced in other matrices, such as starch, as reported by Xiong et al. [79] together with decreased water absorption by the bio-based polymer. SiO_2 nanoparticles are especially interesting due to their ability to add a functional group to their structure, such as a vinyl group, and grafted to a polymer matrix by means of covalent bonds, as reported by Jia et al. in [80]. According to the authors, the nanocomposites prepared by free radical polymerization of PVOH and vinyl SiO_2 presented improved thermal and mechanical properties when compared to the pure PVOH as a consequence of strong covalent bonding between $nSiO_2$ and the polymer matrix. Another study reported that the surface grafting of polymers onto the $nSiO_2$ improved the compatibility of the composites [81], which means that different grafted monomers result in tailored interfacial interactions and enhanced tensile strength.

Other important type of nanoparticles used in food packaging nanocomposite manufacturing encompass metal nanoparticles such as silver nanoparticles (Ag-NPs), due to their well-known antimicrobial activity [82], or bio-based nanoparticles, such as those based on starch nanocrystals and chitin/chitosan. According to Kristo and Biliaderis (2007) [83], the addition of starch nanocrystals to a polymer matrix increases the tensile strength and Young's modulus; however, the elongation at break decreases. Besides, a decrease of around 20% in water vapor permeability was observed in the aforementioned study. Regarding chitin/chitosan nanoparticles, Lu et al. reported that the addition of chitin whiskers to protein films greatly improved the tensile strength and water resistance of the films [81]. Similarly, Sriupayo et al. observed that chitosan films containing chitin nanoparticles presented a general increase on their mechanical properties and water resistance [84].

7.2.2 Methods for the preparation of polymer nanocomposites

There are two main pathways that are followed to prepare polymer nanocomposites. They can be prepared by in situ synthesis of inorganic particles (such as metal oxide nanoparticles); and also by dispersion of fillers (i.e., layered nanoclays) in a polymeric matrix [8].

An appropriate selection of the preparation technique is of major importance for obtaining nanomaterials with desired properties based on a homogeneous distribution of the nanodomains within the continuous polymer matrix [85]. The synthesis of polymer nanocomposites can involve either bottom-up or top-down methodologies. In the bottom-up approach, precursors are the starting point to create well-defined structures from the nanoscale level. Besides block assembly, the strategy refers to the mix of different preformed nanosized units. With already formed nanostructures, these can be hierarchically combined to form the target material [85,86]. In the top-down approach, bulk material is broken down into smaller (sometimes patterned) pieces (nanoparticles), mostly by physical methods. The dispersion of layered nanosized silicates within polymer matrices is the most common example (Fig. 7.3).

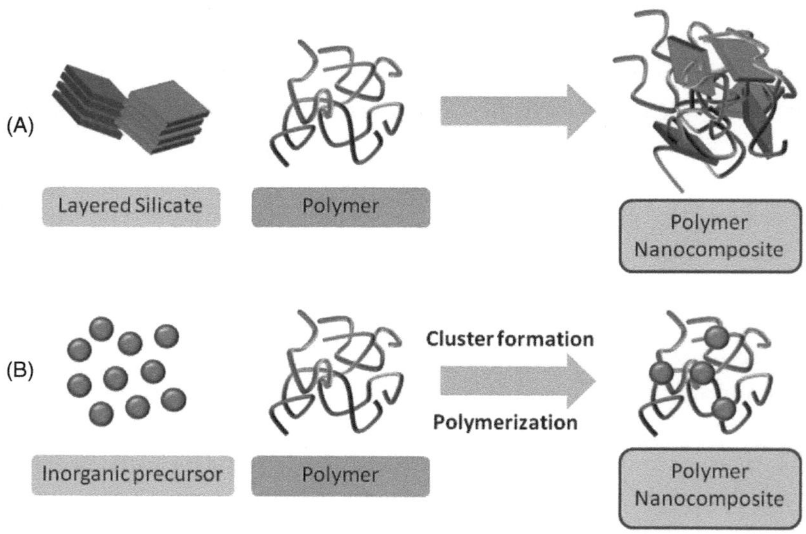

Figure 7.3 Scheme of (A) top-down and (B) bottom-up approaches. *(Adapted from A. Tolentino, Ionic complexes of biodegradable polyelectrolytes, Doctoral dissertation, Universitat Politécnica de Catalunya, Spain, 2014. Available from: http://www.tdx.cat/bitstream/handle/10803/144662/TATC1de1.pdf?sequence=1 [99] with permission).*

Five main routes (1–5) are followed to obtain polymer matrices with dispersed inorganic nanoparticles. The first two use the top-down methodology where nanoparticles are added to the unsolidified polymer matrix, which consists of directly mixing the filler with a polymeric matrix in solution (1) or in melt (2). The bottom-up methodology involves in situ polymerization of monomers in the presence of nanoparticles previously obtained (3) or in situ formation of inorganic particles in the presence of a previously produced polymer (4). Finally, in the last route, inorganic and organic components (polymer matrix) are both formed in situ (5) during the preparation of the composite. The basic methods used to prepare nanocomposites are discussed in Sections 7.2.2.1–7.2.2.5.

7.2.2.1 Preparation from solution

Polymer or prepolymer intercalation from solution, also called exfoliation adsorption, is based on the addition of dispersed polymer to a solution containing layered silicate. The layered silicate is swollen and dispersed in solvent before being mixed with the polymer solution. The polymer chains then intercalate and displace the solvent within the silicate interlayers. Eventually, on removal of the solvent, a multilayer structure forms as the sheets reassemble trapping the polymer chains [88]. This approach is widely used for water-soluble polymers to produce intercalated nanocomposites based on polymers with low or no polarity such as poly(vinyl alcohol), poly(ethylene oxide), poly(vinylpyrrolidone), or poly(acrylic acid) [89]. However, this method is not environmentally friendly given the large amounts of solvents required as only about 10% of the starting material was solid [90]. Emulsion polymerization consists of the polymerization of an organic monomer dispersed in water media. It is considered to be a polymer intercalation method as monomers, usually methyl methacrylate and styrene, are dispersed in water along with emulsifiers, which typically are polyethoxylates (nonionic) or alpha olefin sulfonates (ionic), and added to different silicate concentrations (normally ranging between 1 and 5%) [91]. Finally, the monomer is polymerized with part of the silicate embedded inside the polymer particle and part adsorbed on the particle surface, forming a nanocomposite.

7.2.2.2 Preparation by melt mixing

The melt mixing approach relies on annealing a mixture of polymer and nanoparticles above the softening point of the polymer, typically under shear stress [92]. This process is the most common method of polymer nanocomposite manufacturing when it comes to layered polymer nanocomposites [90]. The mobility of the melted polymer chains enables their diffusion into the galleries present between the host nanolayers during annealing. This makes it possible to obtain an exfoliated distribution of the nanosized material [93].

The polarity difference between the organic polymer matrix and the inorganic nanomaterial can greatly hinder the dispersion thereof, and lead to less favorable formation

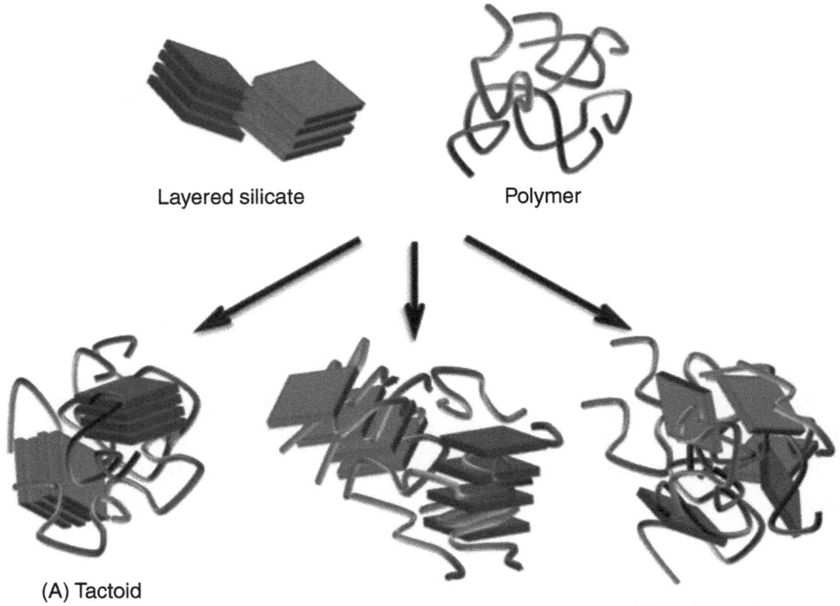

Figure 7.4 Scheme of different layered silicates-polymer composites depending on dispersion effectiveness: (A) tactoid (aggregates), (B) intercalated, and (C) exfoliated (nanocomposites). *(Adapted from A. Tolentino, Ionic complexes of biodegradable polyelectrolytes, Doctoral dissertation, Universitat Politécnica de Catalunya, Spain, 2014. Available from: http://www.tdx.cat/bitstream/handle/10803/144662/TATC1de1.pdf?sequence=1 [99] with permission).*

of intercalated layout or, in the worst case scenario, aggregates (tactoid conformation), as is shown in Fig. 7.4, since the different components tend to remain together with those of same chemical nature. This issue is usually solved by means of functionalization of nanoparticles with polymer compatible organic groups [94]. Unfortunately, the low thermal resistance of the interaction between nanoparticle and the grafted functional groups, together with the high temperatures needed to trigger the melting of a polymer, can degrade the nanomaterials, leading to deficient performance of the final composite. Thus, several factors, such as the nature of the functional groups on the nanoparticles, processing temperature, or applied shear stress, need to be considered when developing the nanocomposite to succeed [95]. The melt mixing process is widely used due to its environmental benignity; its compatibility with a wide variety of polymer matrices with respect to other methodologies that depend on the solubility of the materials or synthesis processes; and compatibility with industrial processes as organic solvent is not needed.

7.2.2.3 Preparation by in situ polymerization

The preparation of nanocomposites by in situ polymerization involves the swelling of the filler in liquid monomer or monomer solution. This swelling occurs as the low-molecular-weight monomer seeps in between the interlayers of the silicate.

Polymerization starts by the application of heat, radiation, initiator diffusion, which triggers the subsequent polyaddition of monomers to obtain the final macrostructure with the nanoparticles embedded within [96]. In this modality, the monomers polymerize in between the interlayers of nanolayered structures, such as clays or graphene, forming intercalated or exfoliated nanocomposites. The advantage of this approach lies in improved exfoliation compared to melt and exfoliation adsorption methods [97].

7.2.2.4 Preparation by in situ synthesis of nanoparticles

The in situ formation of the nanocomposite constituents, based on the bottom-up approach, makes it possible to build well-defined structures, which have different properties from the original precursors. Typically, a polymer acts as a reaction medium where nanoparticles are generated. The desired nanoparticles are then obtained by chemical conversion of their precursor [98]. On the other hand, polymers are robust and chemically stable organic materials. Thus, the resulting nanocomposites exhibit synergetic properties and can be used in applications that could not exist with the polymeric material or the nanoparticles individually. The incorporation of metal nanoparticles into a polymeric matrix can be done using two different means: ex situ or in situ approaches (Fig. 7.5A–C). The inorganic nanoparticles are first synthesized and then introduced in the polymer solution or melt in the ex situ process. This is based on physical entrapment of the metal or metal oxide nanoparticles into the polymer network. However, it is difficult to achieve a homogenous dispersion of the nanoparticles into the polymer matrix. To overcome this difficulty, the in situ approach can be used. Here, the metal or metal oxide particles are generated inside the polymer phase using a metal precursor, which is converted into the desirable nanoparticles. The in situ method allows control over the particle size and the morphology of the particle [98].

7.2.2.5 Preparation by in situ polymerization and inorganic synthesis

The in situ preparation of polymer nanocomposites involves the formation of both polymeric matrix and nanoparticles, which will become embedded in the polymer, at the same time. The most used method for the in situ preparation of polymer nanocomposites is the sol-gel process [100]. This method is related to two reaction steps that lead to the transition from solution state, that is, a colloidal suspension of solid particles in a liquid phase, to a gel state, formed by the interconnected network between phases [101,102]. The chemical reactions involved are: (1) hydrolysis of the precursors to form the reactive groups that will be involved in the subsequent reaction; which consist in (2) the condensation of newly generated reactive groups to form the final network [103,104] (Fig. 7.5C).

7.2.3 Transformation processes of polymer nanocomposites for food packaging

There are several methods to transform polymers and polymer nanocomposites to make final products for food packaging applications. A simple classification can distribute

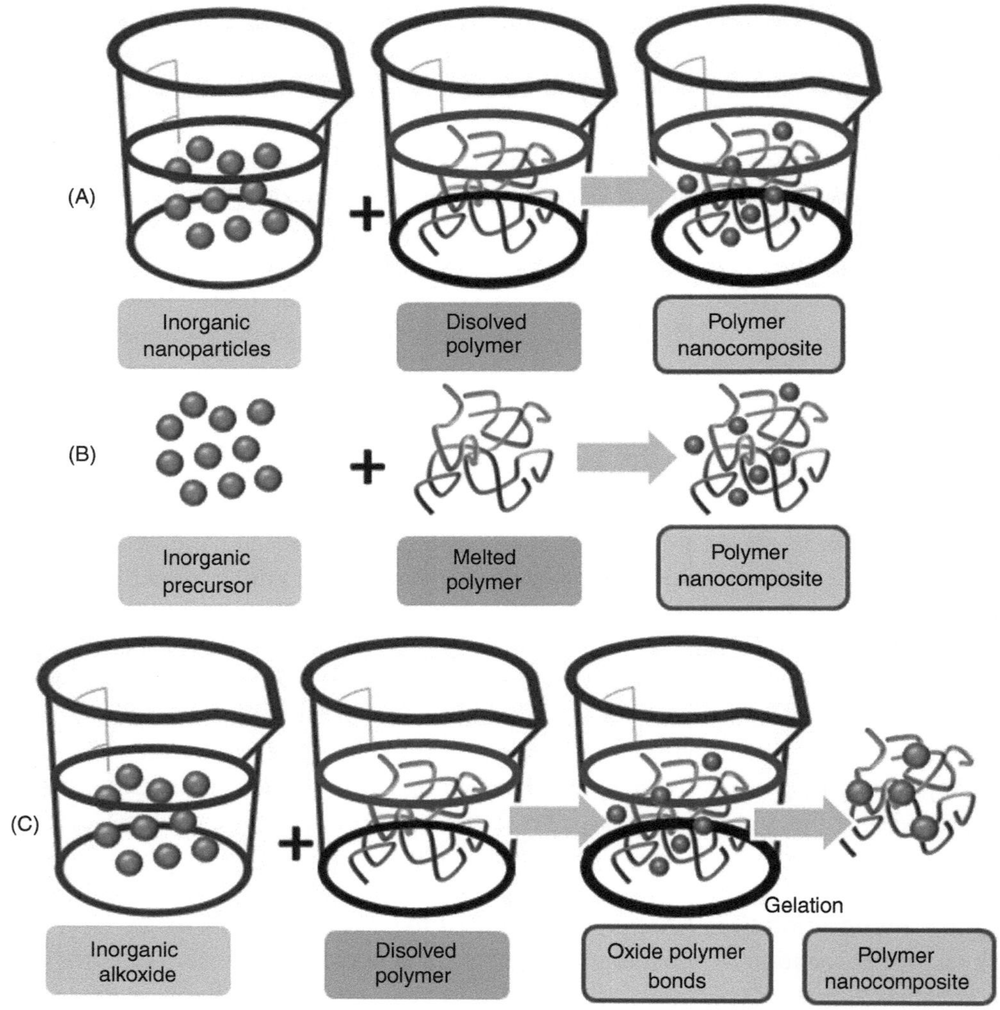

Figure 7.5 Scheme of ex-situ preparation of nanocomposites: (A) solution dispersion, and (B) melt dispersion. Scheme of in situ preparation of nanocomposites: (C) sol-gel process. *(Adapted from A. Tolentino, Ionic complexes of biodegradable polyelectrolytes, Doctoral dissertation, Universitat Politécnica de Catalunya, Spain, 2014. Available from: http://www.tdx.cat/bitstream/handle/10803/144662/TATC1de1.pdf?sequence=1 [99] with permission).*

them into three main preparation procedures based on extrusion and injection processes, as well as their different variants, and thermoforming [105,106] to obtain flat or three-dimensional forms. Extrusion of cast films is used, for instance, for food and textiles packaging, flower wrapping, or photo album page protectors. They are also applied as coating substrates used in extrusion coating processes or laminated to other materials in the formation of more complex films [107,108]. Typically, the transformation of nanocomposites by cast film process involves the use of coextrusion, which is a simultaneous extrusion of two or more materials from a single die to form a multilayered film [109]. In many instances, food packaging applications require the use of films with oxygen barrier capabilities hence films that meet the requirement of a high oxygen barrier material. The poly(ethylene-*co*-vinyl alcohol) copolymer (EVOH) is an example of such materials: it is combined with polyolefin materials in a multilayered structure [110]. The number of layers, their position in the coextrudate, and their individual thickness are all variables that change depending on the particular application of the film [111]. The cooling of the film with cast extrusion is highly efficient and makes possible high production rates [112].

The blown film process is a biaxial stretching of the extruded melt. The molten extrudate passes through an annular die (like a tube) and, when exiting the die, is blown as a bubble. This bubble is cooled by using an air ring along its outer surface. This inflation process will stretch the bubble in the transverse or circumferential direction [113]. During blown film extrusion, melt rheology (both shear and extensional) plays an important role in obtaining a stable process condition. Shear rheology is predominant during the extrusion of the polymer, whereas extensional rheology is predominant when the melt exits from the die and in the process of bubble formation [114]. Therefore, it is useful to understand the shear rheological behavior of the polymer melt to minimize the process difficulties (i.e., high power consumption, lower output, melt fracture). Poor extensional behavior of the polymer melt will also lead to bubble instabilities and inadequate molecular orientation, which will affect the film properties. The most common resin used for blown film production is LDPE due to its high bubble stability and suitability in many packaging applications. LDPE of high molecular weight and long chain branching (LCB) has a substantial impact on the processing behavior [113] and film properties [115].

Another prevailing conventional method for food packaging manufacturing is the injection molding. This process produces plastic parts by injecting material into a mold. The material is fed into a heated barrel, mixed, and forced into a mold cavity, where it cools and hardens to the configuration of the cavity. This technology is used to produce thin-walled plastic parts with high quality, great dimensional accuracy, and productivity for a wide variety of applications. One of the most common usages is plastic housings: a thin-walled enclosure, used in different types of open containers, such as buckets. When hollow objects, such as bottles or jars, are needed in large quantities, a two-stage injection and blow molding process is performed [116]. In a first step a conventional injection molding process is carried out, in which a preform is produced. The preform consists of a fully formed bottle/jar neck with a thick tube of polymer attached, which will form the body. During the blowing step, the preform mold opens and the core rod is rotated and clamped into the hollow, chilled blow mold. The core rod opens and allows compressed air into the preform, which inflates it to the finished article shape. Once the final part is cooled, it is ejected from the mold.

Finally, thermoforming process is the most common option when it comes to manufacturing of trays for food packaging. In this process, a plastic sheet is softened at the heating station and then, it is directed toward the forming station where the forming of the sheet will take place by combining air pressure and male core plugs [117]. Depending on the requirements of the applications, a mono material or a multilayer sheet will be used. Multihead extruders feed into the extrusion die with the differing materials, typically PVC/PE for meat trays, and crystalline PET/amorphous PET for meat trays and ready-to-eat meals [118].

7.3 PROPERTIES OF POLYMER NANOCOMPOSITES FOR FOOD PACKAGING

7.3.1 Polymer nanocomposites with high barrier properties

Polymer nanocomposites can be made into packaging with enhanced gas barrier properties (reduced oxygen, water vapor, and aroma diffusion through the matrix) to maintain food quality and organoleptic properties, hence extending the food shelf life. Gas permeation through polymer nanocomposites follows a two-step mass transport mechanism in which gas molecules are initially adsorbed onto the surface of the polymer and then diffuse through the bulk polymer matrix, where gas can permeate. The incorporation of nanoparticles in the composite, which are typically in the form of layered nanoplatelets, prevents gas permeation [119]. However, the great majority of the studies carried out in this area are focused on the estimation of the diffusion and permeability coefficients, rather than on the estimation of the solubility coefficients, as their contribution is negligible when the stationary state has been attained [120]. The diffusion of gases in the nanocomposite is influenced by the structural parameters of platelets, including the volume fraction, aspect ratio, orientation angle, and, most importantly, the degree of exfoliation in the polymer nanocomposite [121]. The basis of this phenomenon relies on the so-called "tortuous path" (Fig. 7.6). These layered structured materials force gas to travel through the polymer matrix surrounding the particles, which could be for instance nanosilicates, thereby increasing the effective path length [122]. In this way, a simulated case study carried out by Bhunia et al. [123] showed that the total amount of oxygen transfer can be reduced by 97% when platelets with an aspect ratio of 1000 and a volume fraction of 0.07 are dispersed in exfoliated form in the polymer matrix. Gupta et al. [124] reported that polyamide-6 (PA-6)/montmorillonite clay composite (prepared by in situ polymerization) reduced the oxygen transmission rate (OTR) of neat PA from 2.2 to 0.45 cc mil/100 in.^{2} day, when 8% of the silicates were incorporated

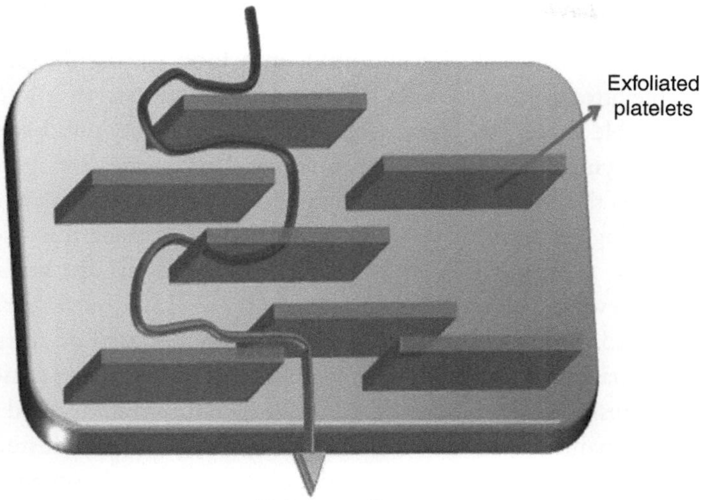

Figure 7.6 Scheme of the "tortuous path" of gas molecules in polymer nanocomposites containing exfoliated platelets.

into the nanocomposites. Akkapeddi et al. found a significant reduction in the OTRs of nylon-6-based nanocomposites in which the silicate plates of high aspect ratio were incorporated [108]. Additionally, OTR was reported to be further reduced by incorporating oxygen scavengers into the matrix. On his behalf, EVOH-kaolinite nanocomposites using a melt intercalation process developed by Cabedo et al. showed that chemical modification of the natural kaolinite helped improve the degree of exfoliation. The treatment also increased the oxygen barrier properties of the nanocomposites, as well as their thermal resistance, glass transition temperature and crystallinity with respect to the neat polymer [125]. De Abreu et al. reported that 5% of layered nanoclay in PP- and LDPE-based composites reduced the OTR from 480 to 374 and from 240 to 210 cc/m^2 per day, respectively [126]. Multilayer films containing montmorillonite layered silicate and polyamide nanocomposite, as oxygen barrier layer, and LDPE, as hydrophobic layer, were developed by Thellen et al. by means of a coextrusion process [127]. Nanofiller content ranging between 3.3 and 3.6% leads to a reduction of the OTR of neat film from 3.7 to 1.1 cc/m^2 per day. Other publications report the development of PET nanocomposites and two different nanoclay platelets using a polymer melt intercalation process [128]. The produced films showed the lowest oxygen permeability with 1 wt.% nanoclay/PET nanocomposite. Meanwhile, Cloisite 15 A (a commercial montmorillonite type layered silicate) exhibited optimal barrier properties, at 2 wt.%. In both cases, the permeability decreased with an increasing degree of exfoliation.

7.3.2 Polymer nanocomposites with antimicrobial properties

The incorporation of antimicrobial compounds into food packaging materials has received considerable attention, as it could help control the growth of microorganisms. It is particularly desirable due to the acceptable structural integrity and barrier properties imparted by the polymeric matrix, and antimicrobial properties by agents embedded within [129]. Due to the high surface-to-volume ratio nanomaterials are able to attach more copies of biological molecules when compared with their microscale counterparts, which leads to a greater efficiency [130]. Nanomaterials are versatile materials that can be used in several ways, such as killing agents [131], antibiotic carriers [132], or growth inhibitors [133]. The quintessential nanocomposites used as antimicrobials for food packaging are based on silver, which is well known for its strong toxicity to a wide range of microorganisms [134], besides its high temperature stability and low volatility [82]. The action mechanism is in line with studies by Kumar and Münstedt, who stated that the antimicrobial activity of silver-based polymers depends on releasing of Ag^+, which binds to electron donor groups in biological molecules containing sulfur, oxygen, or nitrogen [110]. Silver nanocomposites have been prepared and studied by several researchers, including Damm et al. who compared the efficacy of polyamide 6/silver-nano and microcomposites [135]. According to the authors, nanocomposites with a low silver content presented a better increased efficacy against *Escherichia coli* than microcomposites with a much higher silver content. Moreover, polyamide 6 filled with 2 wt.% Ag-NPs was effective against *E. coli*, even after immersed in water for 100 days. Additionally, Li et al. reported that a nanocomposite PE film with Ag-NPs was also able to retard the senescence of jujube, a Chinese fruit [136]. Titanium dioxide (TiO_2) is broadly used as a photocatalytic disinfecting material for surface coatings [137] that has been used to inactivate several food-related pathogenic bacteria [138]. Metal doping, such as silver doping, has been reported to improve visible light absorbance of TiO_2 and increase its photocatalytic activity under UV irradiation, as it was demonstrated by Choi et al. [139]. Even if the great majority of studies are focused on TiO_2/Ag^+ coatings, this combination was used by Cheng et al., who obtained good antibacterial properties from TiO_2/Ag^+ nanoparticles in a nanocomposite with PVC [140]. On the other hand, other antimicrobial mechanisms were proposed by Rabea et al. [141] and Quin et al. [142] based on the chemistry of chitosan nanocomposites containing CNTs. CNTs have been reported to present antibacterial properties as they can puncture microbial cells, causing irreversible damages [143]. Notwithstanding, there are studies addressing the risks associated with their use and these will be further discussed within this chapter.

7.3.3 Polymer nanocomposites with other properties

The improvements of tensile strength and modulus generated by the delaminated nanocomposites structure on polyamide 6–clay hybrids were first reported by the aforementioned studies carried out by Toyota [3]. An increase of 55% of the tensile strength was described, whereas the modulus experimented a rise of 90% with the addition of only 4 wt.% of clay. Studies carried out by Lee and Jang [144] reported that the modulus of intercalated PMMA-clay nanocomposites with 20% weight of clay was found to outperform by 60% to the pristine polymer. It should be pointed out that the contribution of mechanical properties of nanoclay-based nanocomposites both in petrol-based matrices, such as PA, PE, or PET [33,145,146], and bio-based ones, like PLA, PHAs, and starch [33,146–148] are the most studied and successful pathway, especially in the case of exfoliated nanocomposites filled with layered silicates of high aspect ratio. On the other hand, several types of carbon-based nanoparticles have been reported to provide better mechanical strength for packaging applications, such as carbon nanofibers and CNTs [69], although the use for food packaging is still not so clear due to health concerns. Finally, nanocomposites of calcium carbonate [149] or kaolinite [150], typically within polyolefin matrices, have led to an increase in the tensile strength and modulus. The improvement of the mechanical properties is usually closely related to the thermal resistance improvement [151] due to nucleation phenomena in semicrystalline polymers. In this way, increases up to 87°C in the thermal distortion temperature have been achieved in PA/clay nanocomposites [3], together with a thermal expansion coefficient reduction of 45%. In the case of polyolefins, copper nanoparticles have demonstrated to increase the degradation temperature of both LDPE and HDPE due to the increase in thermal conductivity as well as heat capacity of the bulk material [152]. Finally, other promising results have also been reported in terms of chemical resistance [153], flammability [154], optical properties [155], or gas scavenging [156].

7.4 CHARACTERIZATION AND RISK EVALUATION

7.4.1 Characterization of nanocomposites

Characterization tools are of major importance to understand and predict the behavior of polymer nanocomposites as well as their physical and chemical properties. There are several techniques for characterization that have been broadly used in polymer nanocomposite research [157]. The most common ones in terms of structural properties encompass microscopic and X-ray analysis, such as scanning electron microscopy (SEM), transmission electron microscopy (TEM), wide-angle X-ray diffraction (WAXD), and small-angle X-ray scattering (SAXS) [89,158–160]. The SEM provides images of surface features and usually can bring information about the distribution of the different nanoparticles within the polymer matrix. Nevertheless, there are two other microscopies, scanning probe microscopy (SPM) and scanning tunneling microscopy (STM), which are of great utility in nanoparticles research [161]. The SPM is based on the interaction between a sharp tip and a surface to obtain an image. On the other hand, in STM, a sharp conducting tip is held close to a surface (typically about 0.5 nm), so that electrons can go across the gap [162].

This method provides surface structural and electronic information at atomic level. Based on a similar development, the atomic force microscope (AFM) [163] uses a sharp tip to scan across the sample. This method is especially useful to determine mechanical properties of the surface of materials at nanometric scale. TEM allows qualitative understanding of the internal structure, spatial distribution of the various phases, and views of the defective structure through direct visualization, in some cases, of individual atoms [164]. Regarding X-ray-based techniques, WAXD is the most commonly used to probe the nanocomposite structure [165] and occasionally to study the kinetics of the polymer melt intercalation [166] due to its easiness and availability. This technology is especially useful when it comes to layered silicate nanocomposite systems, where a fully exfoliated system is described by the absence of intensity peaks in WAXD pattern [167]. WAXD data have also demonstrated to be comprehensively complemented by TEM [168]. On the other hand, SAXS is typically used to analyze structures on the order of >10 Å. However, X-ray diffraction has found relatively limited success in CNT research [160].

Other common experimental techniques used for the characterization of nanocomposites are nuclear magnetic resonance (NMR). NMR is used in materials where the determination of their chemical structure and interactions, associated with their behavior, gives greater insight into the morphology, surface chemistry and, to a very limited extent, quantification of the exfoliation degree in polymer nanocomposites [169]. Differential scanning calorimetry (DSC) is employed to understand the nature of the changes in the characteristic thermal transitions of the polymer matrix [170]. Fourier transform infrared spectroscopy (FTIR) is used to analyze functional groups on the surface of the nanocomposite and understand better its composition [171]. Dynamic mechanical analysis (DMA) brings information associated with the response of a material to oscillatory deformation as a function of temperature. In DMA, storage modulus is related to elastic response to deformation; loss modulus corresponds to plastic response to deformation; and the ratio of the previous two is an indicator of occurrence of molecular mobility transitions [172]; and resonance Raman spectroscopy is used for complementary structural studies [173]. See Chapter 4 for further information on the analysis of nanomaterials.

7.4.2 Nanosafety aspects related to food contact materials

Currently the potential impacts of engineered nanomaterials on humans and the environment have generated considerable research interest, as their use and diversity of applications in commercial products have grown extensively over the past decade, and continue to grow rapidly. Specifically, the risk that food contact materials containing nanomaterials might pose depends on the (1) likelihood of exposure, which is determined by the amount of substance that migrates from to the material into the food and consequently might be ingested by consumers, and the (2) inherent toxicity of nanomaterials. As detailed in Chapter 5, results from laboratory experiments suggest that some nanomaterials may induce oxidative stress, inflammation, and may bio-accumulate because of their greater ability to pass biological barriers, derived from their different physico–chemical properties compared to their macro size equivalents. However, it is well accepted that there are a large number of deficiencies in the way toxicological studies are designed [174]. It should be also pointed out that interactions between nanomaterials and biological systems are rather complex and greatly depend on nanomaterial size, shape, surface chemistry, and functionalization. This is clearly exemplified by the increased cytotoxicity induced by organoclays in different cell lines, when they are modified with quaternary ammonium cations [175].

Considering the market evolution and nanosafety concerns, the necessity of specific regulations for nanomaterials and nanoenabled products is into debate worldwide. In Europe, despite the fact that no specific legislation on nanotechnology exists, recent regulation specifically addresses nanomaterials which are defined as *natural, incidental, or manufactured material containing particles in an unbound state or as an aggregate or as an agglomerate and where, for 50% or more of the particles in the number size distribution, one or more external dimensions is in the size range 1–100 nm* [176]. Food contact materials are covered by Regulation (EC) 1935/2004, and more specifically nanomaterials are explicitly addressed in the revised "Plastic Food Contact Materials" Regulation 10/2011, where it is stated that only those nanomaterials authorized and mentioned in the Annex I can be used. Regarding the risks of substances in nonplastic materials, such as adhesives, coatings, and inks, the European Commission has planned to launch a consultation on policy options. So far, titanium nitride nanoparticles is the only material authorized as nano in Annex I, defined as material with primary diameter of 20 nm forming aggregates of 100–500 nm. It is also specified in Regulation 10/2011 that migration cannot occur. This differs from other authorized substances to which specific migration or overall migration limits are applied (10 mg/dm^2 of food package surface or 60 mg/kg of food are defined as default values). In addition, two more substances, not listed as nanomaterials, are also included: carbon black and amorphous silica (also currently in use as a food additive in E551, below 100 nm) [177]. Such strict regulation could be the reason for the limited number of available packaging products in the market.

In addition, current recommendations in the USA defined, by the FDA, include guidance on how potential migration to food under worst foreseeable conditions of use should be tested and recommendations on the toxicity experiments that should be carried out are given, for example, mutagenecity in vitro tests. Indeed, the level of data required to support the safety of a food contact material depends on the estimated daily intake. Another consideration to take into account is that migration from materials to food that has been stored under prescribed conditions should overpoise over results obtained using food stimulants. Food stimulants, for example, ethanol, water, acetic acid, vegetable oil at different dilution factors, are assigned according to the hydrophilic/hydrophobic character of the food. Time and temperature are also considered when designing such migration studies. Following this approach, an increasing number of publications have focused on trying to answer the question whether nanomaterials might be released from food contact materials (mostly nanocomposites) by either testing simple immersion protocols defined in the legislation or by using specially designed migration or permeation cells [178]. Literature analysis, including some recent reviews on the topic [179,180], reveals that the most investigated nanomaterials are Ag and clays, followed by ZnO, SiO, TiO_2 or organic materials such as chitosan. Materials are typically embedded in polymeric matrices such as PLA, PE, LDPE, or PET.

Ag nanoparticles are defined as migrating systems because it is well known that their toxicity is mainly caused by Ag ions, which are formed from the oxidation of Ag nanoparticles by air. Ag cations can diffuse out of the matrix and be transported by the moist food environment, where they inhibit bacterial growth. Due to the likelihood of Ag to be ingested by consumers, the scientific community indicates that the incorporation of Ag nanoparticles in materials needs to be regulated to avoid overuse [181]. On the other hand, clays fall within the category of nonmigration systems as they are designed to improve the barrier properties of polymeric materials. Therefore, the property of the nanomaterial is exhibited inside the package, without the requirement of migrating outside packaging layers or into the food. For that reason in terms of the release of these nanomaterials, clay is usually considered as safe. However, some authors have reported low levels of clay constituents (e.g., silica and aluminum) from migration studies in food stimulants [182].

It is widely recognized that the migration of nanomaterials will mainly depend on the nanomaterials physico–chemical properties (i.e., size, functionalization), polymer characteristics, nature of food stimulant, as well as time and temperature. Furthermore, potential migration pathways have been identified: desorption, diffusion, dissolution, and degradation of the matrix [183]. Regarding food contact materials, the main migration routes are diffusion and dissolution. In this regard, it remains a challenge to determine the form in which the nanomaterials are released, which for Ag is not clearly solved yet. For instance, researchers cannot differentiate migration via diffusion mechanism; via release from the food contact material surface, or whether nanomaterials are formed post dissolution, for example, release of Ag^+ and postformation of Ag or $AgCl/Ag_2S$ (nano) particles [184]. Commonly employed analytical techniques are TEM coupled with EDX and ICP-MS, allowing determination of form and concentration of release material, respectively. Analytical methodology is explained in more detail in Chapter 4. Low migration rates as well as miniscule quantities make the detection of nanoparticles in food stimulants difficult to be accomplished. In general terms, it is also evident from the literature that results can often not be compared since first, sample preparation methods and migration tests are frequently adapted and varied in a large number of works. Second, the characterization of the starting materials, including pristine nanomaterials, is generally poor.

In conclusion, enhancing the knowledge on the migration of nanomaterials from food packaging to support reliable risk assessment is necessary, since at present, empirical information that could be used to predict the amount and form of nanomaterials released as a function of their physico–chemical properties, food packaging material, and external conditions, is scarce. This means that more effort should be put into clarifying and evaluating structure–function relationships by systematically modifying compositional parameters. With more data, theoretical modeling of nanomaterials diffusion will gain confidence in the future. Specifically, literature supports the development of new analytical methods able to distinguish between the different potential pathways that can result in the release of nanomaterials, in particular whether nanomaterials leach as ions or single/aggregated/agglomerated nanomaterials. There are still no methods widely accepted to assess particle leaching, but ICP-MS operating in a single-particle mode or the hyphenated techniques such as FFF-ICP-MS in combination with electron microscopy observations will provide advantages over simpler elemental analysis of the food products using ICP-MS.

7.5 OUTLOOK AND CONCLUSION

Nanotechnology will likely encompass, direct or indirectly, almost every feature of the food sector during the upcoming years. The discussed promising applications include nanomaterials as important in packing materials with high gas barrier, antimicrobial properties, and enhanced mechanical and thermal resistance. Nanotechnology is meant to open a wide span of new opportunities within the food packaging sector. However, the introduction of nanomaterials within global public habits still remains uncertain, due to the health and toxicology concerns. As a consequence, consumers are hesitant regarding consuming products involving nanotechnology, even more when it comes to applications related to foods. Specific regulations need to be defined and implemented toward public acceptance of food products that incorporate or utilize nanomaterials. Therefore, the consumption of nanocomposites will be highly dependent on how much trust the public has in industry and the government to protect them from likely hazards. In that sense, the capability of industry and academics to spread their knowledge will play a key role in mitigating public fears about nanorelated food products. Unfortunately, the inherent veil of secrecy of industry toward intellectual property overprotection that could reflect economical return greatly hinders the transparency of their activities. Furthermore, the future of emerging nanotechnologies for food packaging applications is also uncertain upon the way regulatory agencies are able to handle the challenge that nanotechnology represents. Nanotechnology in food packaging strengths must be correctly and delicately counterbalanced against the potential risks of use and abuse of nanomaterials. Furthermore, there are still several gaps of knowledge that need to be filled in terms of product safety, for instance regarding nanomaterial migration through polymer films and further the interaction of nanomaterials with biological functions, as well as the effect of characteristics, concerning size, shape, surface charge, etc. In the same way, a comprehensive definition of identification, quantification, and characterization methods and techniques for nanomaterials in complex packaging structures is also needed. Besides, evaluation of the interrelationship of biodegradability and toxicity of degradation products in new emerging packaging materials such as bio-based materials incorporating nanotechnology needs to be addressed in depth. This chapter aimed to depict a framework of current advances regarding polymer nanocomposites applicable to food packaging to identify the existing gaps and highlight the vast span of possibilities that nanotechnology can bring toward more interactive, smart, and a health-concerned food packaging industry. Although nanocomposites are already well-known materials, their incorporation in food packaging is still in its infancy and the future of this promising technology remains uncertain. Hence, indepth dialog and an open minded attitude will be needed from scientists developing these new approaches, companies in charge of implementing them, as well as the final consumers. If we overtake the existing initial mistrust, nanotechnology will be a key player in a healthier and accessible global food supply.

REFERENCES

[1] Cientifica Report: Nanotechnologies in the food industry, 2006.
[2] A. Okada, Y. Fukushima, M. Kawasumi, S. Inagaki, A. Usuki, S. Sugiyama, O. Kamigaito, Polyamide polymer matrix bonded to silicate, US 4,739,007A (1985).
[3] C. Edser, Auto applications drive commercialization of nanocomposites, Plast. Addit. Compd. 4 (1) (2002) 30–33.
[4] K. Yano, A. Usuki, A. Okada, T. Kurauchi, O. Kamigaito, Synthesis and properties of polyimide–clay hybrid, J. Polym. Sci. 31 (10) (1993) 2493–2498.
[5] E.P. Giannelis, Polymer layered silicate nanocomposites, Adv. Mater. 8 (1) (1996) 29–35.
[6] Y. Kojima, A. Usuki, M. Kawasumi, A. Okada, Y. Fukushima, T. Kurauchi, O. Kamigaito, Mechanical properties of nylon 6–clay hybrid, J. Mater. Res. 8 (5) (1993) 1185–1189.
[7] R. Roy, R.A. Roy, D.M. Roy, Alternative perspectives on "quasi-crystallinity": non-uniformity and nanocomposites, Mater. Lett. 4 (8-9) (1986) 323–328.
[8] G. Kickelbick (Ed.), Hybrid Materials: Synthesis, Characterization, and Applications, Wiley-VCH Verlag GmbH & Co, KGaA, Weinheim, 2007.
[9] F. Hussain, M. Hojjati, M. Okamoto, R.E. Gorga, Polymer-matrix nanocomposites, processing, manufacturing, and application: an overview, J. Compos. Mater. 40 (17) (2006) 1511–1575.
[10] C. Sanchez, P. Belleville, M. Popall, L. Nicole, Applications of advanced hybrid organic–inorganic nanomaterials: from laboratory to market, Chem. Soc. Rev. 40 (2) (2011) 696–753.
[11] G.Z. Cao, Y. Wang, Nanostructures and Nanomaterials: Synthesis, Properties, and Applications, second ed., World Scientific Publisher Co, Singapore, (2011) 596 pp.
[12] P. Judeinstein, C. Sanchez, Hybrid organic–inorganic materials: a land of multidisciplinarity, J. Mater. Chem. 6 (4) (1996) 511–525.
[13] M.Z. Rong, M.Q. Zhang, Y.X. Zheng, H.M. Zeng, K. Friedrich, Improvement of tensile properties of nano-SiO_2/PP composites in relation to percolation mechanism, Polymer 42 (7) (2001) 3301–3304.
[14] X. Liu, Q. Wu, PP/clay nanocomposites prepared by grafting-melt intercalation, Polymer 42 (25) (2001) 10013–10019.
[15] S. Colonna, A. Fina, O. Monticelli, J. Gomez, G. Saracco, Graphite nanoplatelets dispersion by melt reactive extrusion for the preparation of thermally conductive polymer nanocomposites, 2015. Available from: http://graphita.bo.imm.cnr.it/abstracts/Poster_Colonna.pdf
[16] World Packaging Organization (WPO), Market statistics and future trends in global packaging, 2008. Available from: http://www.worldpackaging.org/i4a/doclibrary/index.cfm?category_id=4

[17] J. Lange, Y. Wyser, Recent innovations in barrier technologies for plastic packaging—a review, Packag. Technol. Sci. 16 (4) (2003) 149–158.
[18] S.E. Selke, Plastics in packaging, Plast. Environ. (2003) 137–183.
[19] A.G. Pedroso, D.S. Rosa, Mechanical, thermal and morphological characterization of recycled LDPE/corn starch blends, Carbohydr. Polym. 59 (1) (2005) 1–9.
[20] S. Sahin, P. Yayla, Effects of testing parameters on the mechanical properties of polypropylene random copolymer, Polym. Test. 24 (5) (2005) 613–619.
[21] B.G. Girija, R.R.N. Sailaja, M. Giridhar, Thermal degradation and mechanical properties of PET blends, Polym. Degrad. Stab. 90 (1) (2005) 147–153.
[22] R.B. Pearson, PVC as a food packaging material, Food Chem. 8 (2) (1982) 85–96.
[23] P.H. Daniels, A brief overview of theories of PVC plasticization and methods used to evaluate PVC-plasticizer interaction, J. Vinyl Addit. Technol. 15 (4) (2009) 219–223.
[24] A.R. Berens, Sorption of organic liquids and vapors by rigid PVC, J. Appl. Polym. Sci. 37 (4) (1989) 901–913.
[25] R. Navarro, M. Pérez-Perrino, M. Gómez-Tardajos, H. Reinecke, Phthalate plasticizers covalently bound to PVC: plasticization with suppressed migration, Macromolecules 43 (5) (2010) 2377–2381.
[26] F.E. Bernardin III, G.C. Rutledge, Simulation of mechanical properties of oriented glassy polystyrene, Polymer 48 (24) (2007) 7211–7220.
[27] J.H. Koo, Polymer Nanocomposites: Processing, Characterization, and Applications, McGraw-Hill, New York, (2006).
[28] B. Pukanszky, K. Belina, A. Rockenbauer, F.H.J. Maurer, Effect of nucleation, filler anisotropy and orientation on the properties of PP composites, Composites 25 (3) (1994) 205–214.
[29] Y. Zhong, D. Janes, Y. Zheng, M. Hetzer, D. De Kee, Mechanical and oxygen barrier properties of organoclay-polyethylene nanocomposite films, Polym. Eng. Sci. 47 (7) (2007) 1101–1107.
[30] T.S. Ellis, J.S. D'Angelo, Thermal and mechanical properties of a polypropylene nanocomposite, J. Appl. Polym. Sci. 90 (6) (2003) 1639–1647.
[31] C.I. Calcagno, C.M. Mariani, S.R. Teixeira, R.S. Mauler, The effect of organic modifier of the clay on morphology and crystallization properties of PET nanocomposite, Polymer 48 (4) (2007) 966–974.
[32] A. Arora, G.W. Padua, Review: nanocomposites in food packaging, J. Food Sci. 75 (1) (2010) 43–49.
[33] A. Sorrentino, G. Gorrasi, V. Vittoria, Potential perspectives of bio-nanocomposites for food packaging applications, Trends Food Sci. Technol. 18 (2) (2007) 84–95.
[34] V. Siracusa, P. Rocculi, S. Romani, M.D. Rossa, Biodegradable polymers for food packaging: a review, Trends Food Sci. Technol. 19 (12) (2008) 634–643.
[35] R.P. Babu, K. O'Connor, R. Seeram, Current progress on bio-based polymers and their future trends, Prog. Biomater. 2 (8) (2013) 1–16.
[36] O. Wolf, M. Crank, M. Patel, F. Marscheider-Weidemann, J. Schleich, J. Hüsing, G. Angerer, Techno-economic feasibility of large-scale production of bio-based polymers in Europe. Technical Report EUR, 22103 (2005).
[37] A.M. Clarinval, J. Halleux, Classification of biodegradable polymers, in: R. Smith (Ed.), Biodegradable Polymers for Industrial Applications, Woodhead Publishing Ltd, Cambridge UK, 2005, pp. 3–31.
[38] I. Chandra, R. Rustgi, Biodegradable polymers, Prog. Polym. Sci. 23 (7) (1998) 1273–1335.
[39] J.H. Ha, Antimicrobial food packaging, Novel Food Packag. Tech. (2003) 50–70.
[40] M. Jamshidian, E.A. Tehrany, M. Imran, M. Jacquot, S. Desobry, Poly-lactic acid: production, applications, nanocomposites, and release studies, Compr. Rev. Food Sci. Food Saf. 9 (5) (2010) 552–571.
[41] H. Imran, A.M. Revol-Junelles, A. Martyn, E.A. Tehrany, M. Jacquot, M. Linder, S. Desobry, Active food packaging evolution: transformation from micro- to nanotechnology, Crit. Rev. Food Sci. Nutr. 50 (9) (2010) 799–821.
[42] A. Jiménez, E. Arab-Tehrany, L. Sánchez-González, Progress in biodegradable packaging materials, Future Science e-books, 2014.
[43] N. Peelman, P. Ragaert, B. De Meulenaer, D. Adons, R. Peeters, L. Cardon, F. Devlieghere, Application of bioplastics for food packaging, Trends Food Sci. Technol. 32 (2) (2013) 128–141.
[44] X.Z. Tang, P. Kumar, S. Alavi, K.P. Sandeep, Recent advances in biopolymers and biopolymer-based nanocomposites for food packaging materials, Crit. Rev. Food Sci. Nutr. 52 (5) (2012) 426–442.

[45] T.V. Duncan, Applications of nanotechnology in food packaging and food safety: barrier materials, antimicrobials and sensors, J. Colloid Interface Sci. 363 (1) (2011) 1–24.
[46] P. Frisk, J. Laurent, Nanocomposite polymer container, US Patent 5,876,812A (1999).
[47] B.R Christiani, M. Maxfield, Melt process formation of polymer nanocomposite of exfoliated layered material, US 5,747,560A (1998).
[48] M.V. Burmistr, K.M. Sukhyy, V.V. Shilov, P. Pissis, A. Spanoudaki, I. Sukha, Y.P. Gomza, Synthesis, structure, thermal and mechanical properties of nanocomposites based on linear polymers and layered silicates modified by polymeric quaternary ammonium salts, Polymer 46 (26) (2005) 12226–12232.
[49] A. Galgali, S. Agarwal, A. Lele, Effect of clay orientation on the tensile modulus of polypropylene–nanoclay composites, Polymer 45 (17) (2005) 6059–6069.
[50] R.R. Hedge, Structure and properties of nanoclay reinforced polymer films, fibers and nonwovens, Doctoral dissertation, 2009. Available from: http://trace.tennessee.edu/cgi/viewcontent.cgi?article=1069&context=utk_graddiss
[51] F. Liu, Polymer modified clay minerals: a review, Appl. Clay Sci. 38 (1-2) (2007) 64–76.
[52] J.U. Ha, M. Xanthos, Functionalization of nanoclays with ionic liquids for polypropylene composites, Polym. Compos. 30 (5) (2009) 534–542.
[53] T.D. Fornes, P.J. Yoon, D.L. Hunter, H. Keskkula, D.R. Paul, Effect of organoclay structure on nylon 6 nanocomposite morphology and properties, Polymer 43 (22) (2002) 5915–5933.
[54] J.A.M. Ferreira, P.N.B. Reis, J.D.M. Costa, B.C.H. Richardson, M.O.W. Richardson, A study of the mechanical properties on polypropylene enhanced by surface treated nanoclays, Compos. B 42 (6) (2011) 1366–1372.
[55] S.W. Kim, S.H. Cha, Thermal, mechanical, and gas barrier properties of ethylene–vinyl alcohol copolymer-based nanocomposites for food packaging films: effects of nanoclay loading, J. Appl. Polym. Sci. 131 (11) (2014) 40289–40297.
[56] N. Tenn, N. Follain, J. Soulestin, R. Crétois, S. Bourbigot, S. Marais, Effect of nanoclay hydration on barrier properties of PLA/montmorillonite based nanocomposites, J. Phys. Chem. C 117 (23) (2013) 12117–12135.
[57] M.A.S. Azizi Samir, F. Alloin, A. Dufresne, Review of recent research into cellulosic whiskers, their properties and their application in nanocomposite field, Biomacromolecules 6 (2) (2005) 612–626.
[58] B. Wang, M. Sain, Isolation of nanofibers from soybean source and their reinforcing capability on synthetic polymers, Compos. Sci. Technol. 67 (11-12) (2007) 2521–2527.
[59] M.A.S. Azizi Samir, F. Alloin, J.Y. Sanchez, A. Dufresne, Cellulose nanocrystals reinforced poly(oxyethylene), Polymer 45 (12) (2004) 4149–4157.
[60] M.M.D. Lima, R. Borsali, Rodlike cellulose microcrystals: structure, properties, and applications, Macromol. Rapid Commun. 25 (7) (2004) 771–787.
[61] M.A. Hubbe, O.J. Rojas, L.A. Lucia, M. Sain, Cellulosic nanocomposites: a review, Bioresources 3 (3) (2008) 929–980.
[62] W. Helbert, C.Y. Cavaille, A. Dufresne, Thermoplastic nanocomposites filled with wheat straw cellulose whiskers. Part I: processing and mechanical behavior, Polym. Compos. 17 (4) (1996) 604–611.
[63] P. Podsiadlo, S.Y. Choi, B. Shim, J. Lee, M. Cuddihy, N.A. Kotov, Molecularly engineered nanocomposites: layer-by-layer assembly of cellulose nanocrystals, Biomacromolecules 6 (6) (2005) 2914–2918.
[64] N. Dogan, T.H. McHugh, Effects of microcrystalline cellulose on functional properties of hydroxyl propyl methyl cellulose microcomposite films, J. Food Sci. 72 (1) (2007) E16–E22.
[65] L. Petersson, K. Oksman, Preparation and properties of biopolymer based nanocomposite films using microcrystalline cellulose, in: K. Oksman, M. Sain (Eds.), Cellulose Nanocomposites, Processing, Characterization and Properties. ACS Symposium Series, 938, Oxford University Press, Oxford, 2006, pp. 132–150.
[66] A. Iwatake, M. Nogi, H. Yano, Cellulose nanofiber-reinforced polylactic acid, Compos. Sci. Technol. 68 (9) (2008) 2103–2106.
[67] S.A. Paralikar, J. Simonsen, J. Lombardi, Poly(vinyl alcohol)/cellulose nanocrystal barrier membranes, J. Membr. Sci. 320 (1-2) (2008) 248–258.
[68] M.D. Sanchez-Garcia, E. Gimenez, J.M. Lagaron, Morphology and barrier properties of solvent cast composites of thermoplastic biopolymers and purified cellulose fibers, Carbohydr. Polym. 71 (2) (2008) 235–244.

[69] M.F. De Volder, S.H. Tawfick, R.H. Baughman, A.J. Hart, Carbon nanotubes: present and future commercial applications, Science 339 (6119) (2013) 535–539.
[70] J. Du, H.M. Cheng, The fabrication, properties, and uses of graphene/polymer composites, Macromol. Chem. Phys. 213 (10-11) (2012) 1060–1077.
[71] X. Zhou, E. Shin, K.W. Wang, C.E. Bakis, Interfacial damping characteristics of carbon nanotube-based composites, Compos. Sci. Technol. 64 (15) (2004) 2425–2437.
[72] K.T. Lau, D. Hui, The revolutionary creation of new advanced materials—carbon nanotube composites, Compos. B 33 (4) (2002) 263–277.
[73] J.Y. Kim, S. Han II, S. Hong, Effect of modified carbon nanotube on the properties of aromatic polyester nanocomposites, Polymer 49 (15) (2008) 3335–3345.
[74] Y. Bin, M. Mine, A. Koganemaru, X. Jiang, M. Matsuo, Morphology and mechanical and electrical properties of oriented PVA–VGCF and PVA–MWNT composites, Polymer 47 (4) (2006) 1308–1317.
[75] H. Zeng, C. Gao, Y. Wang, P.C.P. Watts, H. Kong, X. Cui, In situ polymerization approach to multiwalled carbon nanotubes-reinforced nylon 1010 composites: mechanical properties and crystallization behavior, Polymer 47 (1) (2006) 113–122.
[76] S.P. Economopoulos, N. Tagmatarchis, Chemical functionalization of exfoliated graphene, Chem. Eur. J. 19 (39) (2013) 12930–12936.
[77] C. Xiang, P.J. Cox, A. Kukovecz, B. Genorio, D.P. Hashim, Z. Yan, P.M. Sudeep, Functionalized low defect graphene nanoribbons and polyurethane composite film for improved gas barrier and mechanical performances, ACS Nano 7 (11) (2013) 10380–10386.
[78] C.L. Wu, M.Q. Zhang, M.Z. Rong, K. Friedrick, Tensile performance improvement of low nanoparticles filled-polypropylene composites, Compos. Sci. Technol. 62 (10) (2002) 1327–1340.
[79] H.G. Xiong, S.W. Tang, H.L. Tang, P. Zou, The structure and properties of a starch-based biodegradable film, Carbohydr. Polym. 71 (2) (2008) 263–268.
[80] X. Jia, Y. Li, Q. Cheng, S. Zhang, B. Zhang, Preparation and properties of poly(vinyl alcohol)/silica nanocomposites derived from copolymerization of vinyl silica nanoparticles and vinyl acetate, Eur. Polym. J. 43 (4) (2007) 1123–1131.
[81] Y. Lu, L. Weng, L. Zhang, Morphology and properties of soy protein isolate thermoplastics reinforced with chitin whiskers, Biomacromolecules 5 (3) (2004) 1046–1051.
[82] R. Kumar, H. Munstedt, Silver ion release from antimicrobial polyamide/silver composites, Biomaterials 26 (14) (2005) 2081–2088.
[83] E. Kristo, C.G. Biliaderis, Physical properties of starch nanocrystal-reinforced pullulan films, Carbohydr. Polym. 68 (1) (2007) 16–22.
[84] J. Sriupayo, P. Supaphol, J. Blackwell, R. Rujiravanit, Preparation and characterization of a-chitin whisker-reinforced chitosan nanocomposite films with or without heat treatment, Carbohydr. Polym. 62 (2) (2005) 130–136.
[85] E. Ruiz-Hitzky, P. Aranda, M. Darder, M. Ogawa, Hybrid and biohybrid silicate based materials: molecular vs block-assembling bottom-up processes, Chem. Soc. Rev. 40 (2) (2011) 801–828.
[86] G. Kickelbick, Concepts for the incorporation of inorganic building 24. blocks into organic polymers on a nanoscale, Prog. Polym. Sci. 28 (1) (2003) 83–114.
[87] Y.H. Choa, J.K. Yang, B.H. Kim, Y.K. Jeong, J.S. Lee, T. Nakayama, Preparation and characterization of metal: ceramic nanoporous nanocomposite powders, J. Magn. Magn. Mater. 266 (1-2) (2003) 12–19.
[88] F. Hussain, M. Hojjati, M. Okamoto, R.E. Gorga, Polymer-matrix nanocomposites, processing, manufacturing, and application: an overview, J. Compos. Mater. 40 (17) (2006) 1511–1575.
[89] M. Alexandre, P. Dubois, Polymer-layered silicate nanocomposites: preparation, properties and uses of a new class of materials, Mater. Sci. Eng., R 28 (1) (2000) 1–63.
[90] F. Gao, Clay/polymer composites: the story, Mater. Today 7 (11) (2004) 50–55.
[91] D.C. Lee, L.W. Jang, Preparation and characterization of PMMA–clay composite by emulsion polymerization, J. Appl. Polym. Sci. 61 (7) (1996) 1117–1122.
[92] Q.T. Nguyen, D.G. Baird, Preparation of polymer–clay nanocomposites and their properties, Adv. Polym. Tech. 25 (4) (2006) 270–285.
[93] P.H.C. Camargo, K.G. Satyanarayana, F. Wypych, Nanocomposites: synthesis, structure, properties and new application opportunities, Mat. Res. 12 (1) (2009) 1–39.

[94] M. Zanetti, S. Lomakin, G. Camino, Polymer layered silicate nanocomposites, Macromol. Mater. Eng. 279 (1) (2000) 1–9.

[95] W. Lertwimolnun, B.Vergnes, Influence of compatibilizer and processing conditions on the dispersion of nanoclay in a polypropylene matrix, Polymer 46 (10) (2005) 3462–3471.

[96] A. Leszczyńska, J. Njuguna, K. Pielichowski, J.R. Banerjee, Polymer/montmorillonite nanocomposites with improved thermal properties: Part I. Factors influencing thermal stability and mechanisms of thermal stability improvement, Thermochim. Acta 453 (2) (2007) 75–96.

[97] A.P. Kumar, D. Depan, N. Singh Tomer, R.P. Singh, Nanoscale particles for polymer degradation and stabilization—Trends and future perspectives, Prog. Polym. Sci. 34 (6) (2009) 479–515.

[98] S. Sarkar, E. Guibal, F. Quignard, A. SenGupta, Polymer-supported metals and metal oxide nanoparticles: synthesis, characterization, and applications, J. Nanopart. Res. 14 (2) (2012) 1–24.

[99] A. Tolentino, Ionic complexes of biodegradable polyelectrolytes, Doctoral dissertation, Universitat Politécnica de Catalunya, Spain, 2014. Available from: http://www.tdx.cat/bitstream/handle/10803/144662/TATC1de1.pdf?sequence=1

[100] M. Rehahn, Organic/inorganic hybrid polymers, Acta Polym. 49 (5) (1998) 201–224.

[101] V. Maneeratana, Alkoxide-based precursors for direct electrospinning of alumina fibers, University of Florida, 2007. Available from: http://ufdc.ufl.edu/UFE0021743/00001

[102] N.Y. Turova, E.P. Turevskaya, V.G. Kessler, M.I. Yanovskaya (Eds.), The Chemistry of Metal Alkoxides, Kluwer Academic Publishers, Springer, Boston, Dordretch, London, 2002.

[103] C.J. Brinker, G.W. Scherer, Sol-Gel Science: The Physics and Chemistry of Sol-Gel Processing, Academic Press, San Diego, (1990).

[104] U. Schubert, N. Huesing, A. Lorenz, Hybrid inorganic-organic materials by sol-gel processing of organofunctional metal alkoxides, Chem. Mater. 7 (11) (1995) 2010–2027.

[105] Z. Tadmor, C.G. Gogos, Principles of Polymer Processing, second ed., John Wiley & Sons, NJ, (2013).

[106] M.P. Groover, Fundamentals of Modern Manufacturing: Materials Processes, and Systems, third ed., John Wiley & Sons, NJ, (2007).

[107] J.H. Han, Antimicrobial food packaging, in: R. Ahvenainen (Ed.), Novel Food Packaging Techniques, CRC, Woodhead Publishing Limited, Cambridge, 2003, pp. 50–70.

[108] M.K. Akkapeddi, T.J. Kraft, E.P. Socci, US Patent No. 6,423,776. Washington, DC: US Patent and Trademark Office (2002).

[109] P.W.S. Chum, K.C. Koch, US Patent No. 5,089,321. Washington, DC: US Patent and Trademark Office (1992).

[110] D. Granda-Restrepo, E. Peralta, R. Troncoso-Rojas, H. Soto-Valdez, Release of antioxidants from co-extruded active packaging developed for whole milk powder, Int. Dairy J. 19 (8) (2009) 481–488.

[111] G.O. Schroeder, US Patent No. 4,254,169. Washington, DC: US Patent and Trademark Office (1981).

[112] I.M. Ward, Optical and mechanical anisotropy in crystalline polymers, Proc. Phys. Soc. 80 (5) (1962) 1176.

[113] D. Godshall, G. Wilkes, R.K. Krishnaswamy, A.M. Sukhadia, Processing-structure-property investigation of blown HDPE films containing both machine and transverse direction oriented lamellar stacks, Polymer 44 (18) (2003) 5397–5406.

[114] H.H. Winter, M. Mours, Rheology of polymers near liquid-solid transitions in neutron spin echo spectroscopy viscoelasticity rheology, in: Neutron Spin Echo Spectroscopy Viscoelasticity Rheology, Advances in Polymer Science series, vol. 134, Springer, Berlin, Heidelberg, 1997, pp. 165–234.

[115] K.K. Majumder, Blown film extrusion: experimental, modelling and numerical study, Doctoral dissertation, RMIT University, 2008.

[116] D.M. Bryce, Society of Manufacturing Engineers, MI, (1996).

[117] P. Klein, in: A Spanias (Ed.), Fundamentals of Plastics Thermoforming (Synthesis Lectures on Materials Engineering), Morgan & Claypool, San Rafael, CA, 2009, pp. 1–97.

[118] G.L. Robertson, Food Packaging: Principles and Practice, third ed., CRC Press Taylor & Francis Group, Boca Raton, FL, (2012).

[119] G. Choudalakis, A.D. Gotsis, Permeability of polymer/clay nanocomposites: a review, Eur. Polym. J. 45 (4) (2009) 967–984.

[120] A.A. Gusev, H.R. Lusti, Rational design of nanocomposites for barrier applications, Adv. Mater. 13 (21) (2001) 1641–1643.
[121] R.K. Bharadwaj, Modeling the barrier properties of polymer-layered silicate nanocomposites, Macromolecules 34 (26) (2001) 9189–9192.
[122] C. Silvestre, D. Duraccio, S. Cimmino, Food packaging based on polymer nanomaterials, Prog. Polym. Sci. 36 (12) (2011) 1766–1782.
[123] K. Bhunia, S. Dhawan, S.S. Sablani, Modeling the oxygen diffusion of nanocomposite-based food packaging films, J. Food Sci. 77 (7) (2012) 29–38.
[124] B. Gupta, M.F. Lacrampe, P. Krawczak, Polyamide-6/clay nanocomposites: a critical review, Polym. Compos. 14 (1) (2006) 13–38.
[125] L. Cabedo, E. Gimenez, J.M. Lagaron, R. Gavara, J.J. Saura, Development of EVOH-kaolinite nanocomposites, Polymer 45 (15) (2004) 5233–5238.
[126] D.A. Pereira de Abreu, P. Parseiro Losada, I. Angulo, J.M. Cruz, Development of new polyolefin films with nanoclays for application in food packaging, Eur. Polym. J. 43 (6) (2007) 2229–2243.
[127] C. Thellen, S. Schirmer, J.A. Ratto, B. Finnigan, D. Schmidt, Co-extrusion of multilayer poly(m-xylylene adipimide) nanocomposite films of high oxygen barrier packaging applications, J. Membr. Sci. 340 (1) (2009) 45–51.
[128] M. Frounchi, A. Dourbash, Oxygen barrier properties of poly(ethylene terephthalate) nanocomposites films, Macromol. Mater. Eng. 294 (1) (2009) 68–74.
[129] J.W. Rhim, P.K. Ng, Natural biopolymer-based nanocomposite films for packaging applications, Crit. Rev. Food Sci. Nutr. 47 (4) (2007) 411–433.
[130] P.G. Luo, F.J. Stutzenberger, Nanotechnology in the detection and control of microorganisms, Adv. Appl. Microbiol. 63 (2) (2008) 145–181.
[131] P.K. Stoimenov, R.L. Klinger, G.L. Marchin, K.J. Klabunde, Metal oxide nanoparticles as bactericidal agents, Langmuir 18 (17) (2002) 6679–6686.
[132] H. Gu, P.L. Ho, E. Tong, L. Wang, B. Xu, Presenting vancomycin on nanoparticles to enhance antimicrobial activities, Nano Lett. 3 (9) (2003) 1261–1263.
[133] N. Cioffi, L. Torsi, N. Ditaranto, G. Tantillo, L. Ghibelli, L. Sabbatini, E. Traversa, Copper nanoparticle/polymer composites with antifungal and bacteriostatic properties, Chem. Mater. 17 (21) (2005) 5255–5262.
[134] S.Y. Liau, D.C. Read, W.J. Pugh, J.R. Furr, A.D. Russell, Interaction of silver nitrate with readily identifiable groups: relationship to the antibacterialaction of silver ions, Lett. Appl. Microbiol. 25 (4) (1997) 279–283.
[135] C. Damm, H. Münstedt, A. Rösch, The antimicrobial efficacy of polyamide 6/silver-nano-and microcomposites, Mater. Chem. Phys. 108 (1) (2008) 61–66.
[136] H. Li, F. Li, L. Wang, J. Sheng, Z. Xin, L. Zhao, Q. Hu, Effect of nano-packing on preservation quality of Chinese jujube, Food Chem. 114 (2) (2009) 547–552.
[137] A. Fujishima, T.N. Rao, D.A. Tryk, Titanium dioxide photocatalysis, J. Photochem. Photobiol., C 1 (1) (2000) 1–21.
[138] J.M. Robertson, P.K. Robertson, L.A. Lawton, A comparison of the effectiveness of TiO_2 photocatalysis and UVA photolysis for the destruction of three pathogenic micro-organisms, J. Photochem. Photobiol., A 175 (1) (2005) 51–56.
[139] W. Choi, A. Termin, M.R. Hoffmann, The role of metal ion dopants in quantum-sized TiO_2: correlation between photoreactivity and charge carrier recombination dynamics, J. Phys. Chem. 98 (51) (1994) 13669–13679.
[140] Q. Cheng, C. Li, V. Pavlinek, P. Saha, H. Wang, Surface-modified antibacterial TiO_2/Ag^+ nanoparticles: preparation and properties, Appl. Surf. Sci. 252 (12) (2006) 4154–4160.
[141] M.E. Badawy, E.I. Rabea, T.M. Rogge, C.V. Stevens, G. Smagghe, W. Steurbaut, M. Höfte, Synthesis and fungicidal activity of new N, O-acyl chitosan derivatives, Biomacromolecules 5 (2) (2004) 589–595.
[142] C. Qin, H. Li, Q. Xiao, Y. Liu, J. Zhu, Y. Du, Water-solubility of chitosan and its antimicrobial activity, Carbohydr. Polym. 63 (3) (2006) 367–374.
[143] S. Kang, M. Pinault, L.D. Pfefferle, M. Elimelech, Single-walled carbon nanotubes exhibit strong antimicrobial activity, Langmuir 23 (17) (2007) 8670–8673.

[144] D.C. Lee, L.W. Jang, Preparation and characterization of PMMA clay hybrid composite by emulsion polymerization, J. Appl. Polym. Sci. 61 (7) (1996) 1117–1122.
[145] Q. Chaudhry, M. Scotter, J. Blackburn, R. Bryony, A. Boxall, L. Castle, R. Aitken, R. Watkin, Applications and implications of nanotechnologies for the food sector, Food Addit. Contam. A 25 (3) (2008) 241–258.
[146] M.D. Sanchez-Garcia, A. Lopez-Rubio, J.M. Lagaron, Natural micro and nanobiocomposites with enhanced barrier properties and novel functionalities for food biopackaging applications, Trends Food Sci. Technol. 21 (11) (2010) 528–536.
[147] C. Zhijiang, Y. Guang, Optical nanocomposites prepared by incorporating bacterial cellulose nanofibrils into poly(3-hydroxybutyrate), Mater. Lett. 65 (2) (2011) 182–184.
[148] M. Avella, J.J. De Vlieger, M.E. Errico, S. Fischer, P. Vacca, M.G. Volpe, Biodegradable starch/clay nanocomposite films for food packaging applications, Food Chem. 93 (3) (2005) 467–474.
[149] M. Avella, S. Cosco, M.L.D. Lorenzo, E.D. Pace, M.E. Errico, G. Gentile, iPP based nanocomposites filled with calcium carbonate nanoparticles: structure/properties relationships, Macromol. Symp. 234 (1) (2006) 156–162.
[150] M.P. Villanueva, L. Cabedo, J.M. Lagarón, E. Giménez, Comparative study of nanocomposites of polyolefin compatibilizers containing kaolinite and montmorillonite organoclays, J. Appl. Polym. Sci. 115 (3) (2010) 1325–1335.
[151] M. Avella, S. Cosco, M.L. Di Lorenzo, E. Di Pace, M.E. Errico, Influence of $CaCO_3$ nanoparticles shape on thermal and crystallization behavior of isotactic polypropylene based nanocomposites, J. Therm. Anal. Calorim. 80 (1) (2005) 131–136.
[152] J.A. Molefi, A.S. Luyt, I. Krupa, Comparison of the influence of Cu micro- and nano-particles on the thermal properties of polyethylene/Cu composites, eXPRESS Polym. Lett. 3 (10) (2009) 639–649.
[153] J.P. Harcup, A.F. Yee, M.K. Akkapedi, Deformation and fracture of polymer silicate layer nanocomposites, in: Society of Plastics Engineers (Ed.), vol. 3, Proceedings of the Conference ANTEC'99, 2–6 May 1999, New York City, pp. 3396–3398.
[154] J.W. Gilman, T. Kashiwagi, J.D. Lichtenhan, Nanocomposites: a revolutionary new flame retardant approach, Sampe J. 33 (1997) 40–46.
[155] Z. Gao, W. Xie, J.M. Hwu, L. Wells, W.P. Pan, The characterization of organic modified montmorillonite and its filled PMMA nanocomposites, J. Therm. Anal. Calorim. 64 (2) (2001) 467–475.
[156] L. Xiao-e, A.N. Green, S.A. Haque, A. Mills, J.R. Durrant, Light-driven oxygen scavenging by titania/polymer nanocomposite films, J. Photochem. Photobiol., A 162 (2) (2004) 253–259.
[157] M. Meyyappan, Carbon Nanotubes: Science and Applications, CRC Press, Boca Raton, FL, (2004).
[158] C. Park, O.O. Park, J.G. Lim, H.J. Kim, The fabrication of syndiotactic polystyrene/organophilic clay nanocomposites and their properties, Polymer 42 (17) (2001) 7465–7475.
[159] O.O. Christopher, M. Lerner, , Nanocomposites and Intercalation Compound, Encyclopedia of Physical Science and Technology, vol. 10, Academic Press, San Diego, 2001.
[160] R.E. Gorga, R.E. Cohen, Toughness enhancements in poly(methyl methacrylate) by addition of oriented multiwall carbon nanotubes, J. Polym. Sci. B 42 (14) (2004) 2690–2702.
[161] A.K. Barick, D.K. Tripathy, Preparation and characterization of carbon nanofiber reinforced thermoplastic polyurethane nanocomposites, J. Appl. Polym. Sci. 124 (1) (2012) 765–780.
[162] The Royal Society & The Royal Academy of Engineering, Nanoscience and Nanotechnologies: opportunities and uncertainties, 2004. Available from: https://royalsociety.org/~/media/Royal_Society_Content/policy/publications/2004/9693.pdf->
[163] X. Li, W.C. Chang, Y.J. Chao, R. Wang, M. Chang, Nanoscale structural and mechanical characterization of a natural nanocomposite material: the shell of red abalone, Nano Lett. 4 (4) (2004) 613–617.
[164] A.B. Morgan, J.W. Gilman, Characterization of polymer-layered silicate (clay) nanocomposites by transmission electron microscopy and X-ray diffraction: a comparative study, J. Appl. Polym. Sci. 87 (8) (2003) 1329–1338.
[165] P. Reichert, J. Kressler, R. Thomann, R. Mulhaupt, G. Stoppelmann, Nanocomposites based on a synthetic layer silicate and polyamide-12, Acta Polym. 49 (2-3) (1998) 116–123.
[166] K. Yano, A. Usuki, Synthesis and properties of polyimide–clay hybrid, J. Polym. Sci. A 31 (10) (1993) 2493–2498.

[167] J.H. Park, S. Jana, Mechanism of exfoliation of nanoclay particles in epoxy-clay nanocomposites, Macromolecules 36 (8) (2003) 2758–2768.
[168] T.K. Chen, Y.I. Tien, K.H. Wei, Synthesis and characterization of novel segmented polyurethane/clay nanocomposites via poly (ε-caprolactone)/clay, J. Polym. Sci. A 37 (13) (1999) 2225–2233.
[169] D.L. Van der Hart, A. Asano, J.W. Gilman, NMR measurements related to clay-dispersion quality and organic-modifier stability in nylon-6/clay nanocomposites, Macromolecules 34 (12) (2001) 3819–3822.
[170] T.D. Fornes, D.R. Paul, Crystallization behavior of nylon 6 nanocomposites, Polymer 44 (14) (2003) 3945–3961.
[171] S.S. Ray, M. Okamoto, Polymer/layered silicate nanocomposites: a review from preparation to processing, Prog. Polym. Sci. 28 (11) (2003) 1539–1641.
[172] X. Fu, S. Qutubuddin, Polymer–clay nanocomposites: exfoliation of organophilic montmorillonite nanolayers in polystyrene, Polymer 42 (2) (2001) 807–813.
[173] R. Haggenmueller, H.H. Gommans, A.G. Rinzler, J.E. Fischer, K.I. Winey, Aligned single-wall carbon nanotubes in composites by melt processing methods, Chem. Phys. Lett. 330 (3) (2000) 219–225.
[174] H.F. Krug, Nanosafety research—are we on the right track?, Angew. Chem. Int. Ed. 53 (46) (2014) 12304–12319.
[175] G. Janer, E. Fernández-Rosas, E. Mas Del Molino, D. González-Gálvez, G. Vilar, C. López-Iglesias, V. Ermini, S. Vázquez-Campos, In vitro toxicity of functionalised nanoclays is mainly driven by the presence of organic modifiers, Nanotoxicology 8 (3) (2014) 279–294.
[176] European Commission, Commission Recommendation of 18 October 2011 on the definition of nanomaterial, Off. J. Eur. Union (2011) 2011/696/EU: 38–40.
[177] H. Bouwmeester, P. Brandhoff, H.J.P. Marvin, S. Weigel, R.J.B. Peters, State of the safety assessment and current use of nanomaterials in food and food production, Trends Food Sci. Technol. 40 (2) (2014) 200–210.
[178] Y. Song, J. Koontz, R. Juskelis, Y. Zhao, Static liquid permeation cell method for determining the migration parameters of low molecular weight organic compounds in polyethylene terephthalate, Food Addit. Contam. A 30 (10) (2013) 1837–1848.
[179] J.C. Hannon, J. Kerry, M. Cruz-Romero, M. Morris, E. Cummins, Advances and challenges for the use of engineered nanoparticles in food contact materials, Trends Food Sci. Technol. 43 (1) (2015) 43–62.
[180] K.K. Kuorwel, M.J. Cran, J.D. Orbell, S. Buddhadasa, S.W. Bigger, Review of mechanical properties, migration, and potential applications in active food packaging systems containing nanoclays and nanosilver, Compr. Rev. Food Sci. Food Saf. 14 (4) (2015) 411–430.
[181] B. Reidy, A. Haase, A. Luch, K. Dawson, I. Lynch, Mechanisms of silver nanoparticle release, transformation and toxicity: a critical review of current knowledge and recommendations for future studies and applications, Materials 6 (6) (2013) 2295–3350.
[182] M. Farhoodi, S.M. Mousavi, R. Sotudeh-Gharebagh, Z. Emam-Djomeh, A. Oromiehie, Migration of aluminum and silicon from PET/clay nanocomposite bottles into acidic food stimulant, Packag. Technol. Sci. 27 (2) (2014) 161–168.
[183] G.O. Noonan, A.J. Whelton, D. Carlander, T.V. Duncan, Measurement methods to evaluate engineered nanomaterial release from food contact materials, Compr. Rev. Food Sci. Food Saf. 13 (4) (2014) 679–692.
[184] C. Levard, B.C. Reinsch, F.M. Michel, C. Oumahi, G.V. Lowry, G.E. Brown Jr., Sulfidation processes of PVP-coated silver nanoparticles in aqueous solution: impact on dissolution rate, Environ. Sci. Technol. 45 (12) (2011) 5260–5266.

CHAPTER EIGHT

Coatings and Inks for Food Packaging Including Nanomaterials

Lorenzo Bautista*, Laura Molina*, Sandra Niembro*, José Manuel García*,, Jose López*, Alejandro Vílchez***
*LEITAT Technological Centre, Terrassa, Spain
**University of Barcelona, Barcelona, Spain

Contents

8.1 INTRODUCTION

Coatings and printed inks are usually applied for improving functionality and performance of food packaging materials. Those including nanomaterials (or nanobased coatings and inks) are developed and validated to achieve novel and improved performance compared to conventional materials for food packaging applications. Very high surface area per weight ratio, together with specific functionalities derived from the nanosize, are key aspects to understand the increased interest in research on nanobased coatings and inks for food packaging applications.

Emerging Nanotechnologies in Food Science

New or improved properties can be achieved when functional nanomaterials are embedded in packaging materials, such as nanocomposites, or in the form of nanobased coatings and inks. Nanocomposites can be efficient, durable, and easily manufactured in a single compounding step (see Chapter 7 for further details). However, when nanocomposites incorporate nanomaterials as part of their internal structure, and this can be done, for example, by extrusion or injection molding, the amount of nanomaterials required is much higher than when applying them as nanobased coatings and inks. This implies that nanocomposites can be more expensive than nanobased coated/printed materials, the total packaging costs of the former being expected around 10% of the food product cost [1]. Besides, the incorporation of functionalities on the surface of materials by coatings or inks is sometimes independent of the type of packaging (i.e., thermolabile materials); therefore this type of processing is much more versatile than those associated with nanocomposites.

8.2 TYPES OF COATINGS/INKS FOR FOOD PACKAGING INCLUDING NANOMATERIALS

8.2.1 Main functionalities

The main functionalities in materials requested by the food packaging industry that are conferred by nanobased coating are mechanical, barrier, electrical/electronic, optical, and antimicrobial properties. A range of examples showing how solutions involving nanotechnology are helping to meet the industry needs are exposed as follows:

1. Abrasion and scratch resistance food packaging materials with improved mechanical resistance such as glass bottles, or ultrahigh molecular weight polyethylene used in severe operating conditions in food processing can be prepared by incorporating coatings containing ceramic [i.e., silica (SiO_2)] nanoparticles.
2. Food-grade coatings with high anticorrosive barrier are required for food or beverage cans, particularly for steel cans. Such coatings can be developed by applying polymeric coatings containing nanocarriers that can confer improved anticorrosive protection.
3. Vacuum-coated ceramic nanolayers (i.e., SiO_x) are currently used to improve gas and water vapor barrier properties in flexible food packaging materials.
4. Conductive and semiconductive printed inks based on metallic nanoparticles [e.g., silver (Ag), copper (Cu)] are increasingly used as part of gas and temperature sensors for food packaging applications.
5. Titanium dioxide (TiO_2) nanoparticles are used in food packaging materials due to their high UV absorption and photocatalytic properties.
6. Effective and durable antimicrobial coatings based on nanocapsules made from natural polymers.

8.2.2 Mechanical properties

Elongation at break and tensile strength are both important characteristics for packaging materials. In general, these properties are inversely related. Hence, when a particular material presents high elongation at break, its tensile strength is usually low. And the opposite is true; a material with high tensile strength presents low elongation at break. The presence of nanomaterials in inks and coatings can enhance the mechanical properties of the packaging. Usually, the tensile strength of food packaging increases with small amounts of nanoparticles included in nanocomposite coatings; however this is followed by a decrease in the tensile strength when the loading of nanoparticles increases. The elongation at rupture is also greatly affected by the inclusion of nanoparticles: this property decreases as a result of nanoparticle agglomeration, which hinders the intercalation or exfoliation of the nanocomposite coatings.

The food packaging industry is increasingly interested in bio-based coatings composed of starch, chitosan, celluloses, lipids, and/or proteins, among others, including nanoparticles (i.e., biocomposite coatings) because they can improve the properties of plastic substrates without jeopardizing its original attributes and optimize their cost efficiency [2]. For instance, the tensile strength of a protein-based biopolymer such as wheat gluten coating deposited on paper increases with the addition of low amounts of cellulose nanocrystals (CNC) and TiO_2 nanoparticles. However, higher amounts of these nanoadditives decrease the resistance of the polymeric film against fracture [3]. Similar behavior has been obtained for polylactic acid (PLA) films coated with nanoclay inks [4]. Other bio-based nanocomposite coatings composed of chitosan and starch can improve their mechanical properties after addition of magnetite (Fe_3O_4) nanoparticles. The addition of Fe_3O_4 nanoparticles to the film (up to 10%) increases its tensile strength while decreasing the elongation at break [5]. The development of bio-based nanocomposite coatings has aimed at improving not only the mechanical properties of the polymeric resin (i.e., stiffness, impact, wear resistance), but also their flame retardant behavior and their barrier properties against gases, vapors, and radiation [6]. Nevertheless, insignificant effect on the mechanical properties of film was observed when nanoparticles were coated directly onto film, instead of being filled into the polymeric matrix. For instance, mechanical properties of zinc oxide (ZnO) coated PVC films [7], and cellulose nanocrystals coated flexible food packaging films [8], are similar to bare films. This confirms that the improvement of the mechanical properties is not related to the nanoparticles themselves but to the effect of their interaction with polymeric matrices in the nanocomposites.

Abrasion and scratch resistance of food packaging, which are other main mechanical properties in food packaging, can be enhanced by using varnishes containing ceramic nanoparticles. An example of such coating formulation is an UV-cured over-print varnish containing LP X 20470 (BYK-Chemie), which is a colloidal SiO_2 nanoparticle dispersion in tripropylene glycol diacrylate [9].

8.2.3 Barrier properties

Besides polymeric nanocomposites with enhanced barrier properties (covered in Chapter 7), nanobased coatings with the same use are commercially available in food packaging applications. Barrier properties can result from the following phenomena: (1) hydrophobic character given by barrier coating materials (i.e., nanoclays, polyolefines, perfluorinated polymers); (2) reduction of the free volume of the polymeric matrix; (3) possible changes in the orientation of polymeric crystals within the matrix; and (4) reduction of the diffusion of gases and vapors across the thickness of the coating materials due to their "tortuous path" across the packaging material, which has been discussed and displayed in Chapter 7. The enhanced barrier properties thus result from the higher aspect ratio associated with the distribution of nanolayered materials dispersed in a polymer matrix compared to conventional random distribution of particles in composites [10,11].

The most important nanobased barrier coatings are ceramic nanocoatings, layered nanoclays, sol–gel coatings, and bio-nanocomposite coatings. Ceramic nanocoatings obtained through homogeneous deposition of silicon oxide (i.e., SiO_x) nanolayers (40–80 nm onto PET, PA, PP, and PLA) by plasma processing [12] or electron beam evaporation [13] are currently commercialized by the company Amcor. SiO_x barrier nanocoatings are typically used in flexible packaging applications like stand-up pouches, pillow pouches, sachets, flow-wraps, lidding films for trays and cups, tube laminates, and others. They constitute an environment friendly alternative to chlorine-containing PVDC lacquered films. SiO_x barrier nanocoatings are applied in different laminated structures and each structure has its own barrier. Fig. 8.1 shows oxygen and moisture barrier properties of most commonly used laminates in food packaging.

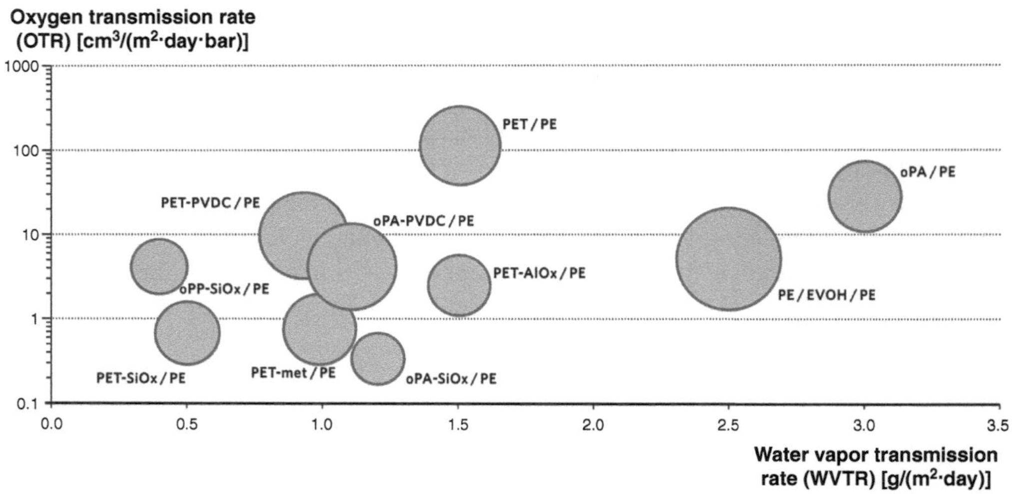

Figure 8.1 ***Barrier properties of different laminated structures for food packaging.*** *(Adapted from Amcor, Ceramis, 2016. Available from: http://www.amcor.com [13]).*

As it can be seen in Fig. 8.1, the barrier caused by SiO_x does not decrease under high humidity conditions, as it is the case with ethylene vinyl alcohol (EVOH). This technology is able to achieve very low oxygen transmission rates (OTR) and water vapor transmission rates (WVTR), which are potentially useful for applications even in the encapsulation of organic light emitting diodes, which is a very demanding application [14].

The cost of vacuum technologies for the deposition of nanocoatings is relatively high. At present, there is no other alternative technology that can achieve high barrier level with high transparency and low haze obtained when using SiO_x nanolayers. Novel acrylic nanocoatings are developed to offer an alternative to SiO_x nanolayers. These are prepared with environment friendly and low cost atmospheric pressure plasma enhanced chemical vapor deposition (AP-PECVD) [15]. Such nanocoatings are highly flexible and, hence, potentially useful for improving performance of current high barrier ceramic nanocoatings. Other technologies based on wet chemistry formulations including nanomaterials (i.e., layered nanoclays, sol–gel, and (bio)nanocomposite coatings) have also been developed. In particular, layered inorganic silicates such as nanoclays have attracted great attention by the packaging industry due to their environmental friendliness, natural abundance, low cost, simple processability, and significant improvement of barrier properties [16]. Layered silicates present approximately 1 nm thickness; layer diameter of 100–1000 nm; 6000–10,000 layers per particle; particle size of 6–13 μm; 700–800 m^2/g surface area; negative surface charge [17,18], and the possibility of presenting different structures of layered nanocomposites. The high surface area of nanoclays intercalated or exfoliated (preferably) in polymeric matrices; their nanolayered structures and hydrophobic nature explain the improvements of mechanical, barrier, chemical resistance, and flame retardant properties associated with their use [19]. It was reported that, with only 1% (wt.) of the nanoclay Cloisite 30B contained in gravure inks, OTR improved from 1600 $cm^3/m^2 \cdot day$ to 1000 $cm^3/m^2 \cdot day$, and modified nanoclays present better performance than nonmodified ones [19].

Several nanoclay-based products have been recently commercialized as part of plastic composites and barrier coatings for food packaging applications [20]. Durethan KU2-2601, from Nanocor, is based on polyamide 6 and montmorillonite composites. These provide excellent gas and moisture barrier properties, strength, toughness, abrasion, and chemical resistance [21]. Imperm products, from Nanocor, Voridian and Mitsubishi Gas Chemical Company, are based on semiaromatic polyamide-based nanocomposites that extend the shelf life of packaged food (e.g., multilayer PET bottles) [22]. Aegis OXCE, from Honeywell Polymer, is a polyamide 6-nanoclay composite modified with an additional oxygen scavenger used for high-oxygen barrier packaging for beer and flavored alcoholic beverages. Triton Systems, Inc., the US military and NASA have developed together (EVOH)-nanoclay composites for long-shelf life packaging (i.e., nonrefrigerated food). Nanolok, from InMat Inc., is a high barrier, water-based nanocomposite coating for transparent packaging applications with suitable oxygen barrier, up to 80% relative humidity, and more cost effective than EVOH [23]. Nano-Seal, from

NanoPack Inc., is a water-based coating made of PVOH and VMT with good oxygen and aroma barrier properties [24]. Nanoblend, from PolyOne Corp., is a homo- and modified polypropylene (PP), linear low-density polyethylene (LLDPE), low-density polyethylene (LDPE), high density polyethylene (HDPE), or an ethylene copolymer matrix-based composite with 40% nanoclays. Polyolefin-based nanocomposites provide higher barrier properties but poorer optical properties (e.g., haze and clarity) than polyamide-based ones [25]. Finally, Plantic R1 Tray, from Plantic Technologies Ltd., is a thermoformed starch-based nanoclay composite that can be used as a bio-based barrier coating [26].

Sol–gel coatings are another important solution with barrier properties. These are based on hydrolysis, condensation, and polycondensation reactions of metal alkoxides obtained under mild operations conditions. Sol–gel coatings present high thermal stability, mechanical strength, and gas barrier properties [27,28]. These are applied to polar polymeric materials such as PET [29] and PA-6 [30]. However, lack of adhesion and cracking is observed when applying these coatings onto polyolefins [31]; to mitigate this, surface modifiers, such as silanes, and/or surface pretreatments, such as atmospheric pressure plasma, are required in sol–gel coating formulations. Poly(lactic acid) (PLA) is an abundant biodegradable polymer with poor barrier properties. In contrast, hybrid organic–inorganic (PLA)/silica (SiO_2) coatings increase the barrier properties of the starting polymer in food packaging applications [32]. Other PLA/SiO_2 hybrids applied by bar coater (or casting) have also been recently investigated. The transparency and optical transmittance of films were over 92%, and OTR and WVTR properties were increased by 69.7 and 45.7%, respectively, over those of neat PLA films [33]. Finally, a plasma posttreatment applied onto hybrid inorganic–organic coatings can increase the inorganic character of the outermost layer, which can improve their wetting, adhesion, and gas barrier properties [34].

In addition, many of the current food packaging materials with high barrier properties are based on multilayer systems (e.g., laminated and/or coextruded materials) including usually PE thermo-sealable outer layers; EVOH (as high barrier layer); and other layers made of PET or PA and adhesives. These multilayered gas barrier materials are effective but they are not recyclable. They can present high thickness which makes the final product less light and more expensive.

Bio-nanocomposite coatings consist of a single monolayer containing a high barrier nanomaterials well dispersed in a bio-polymeric matrix. This type of solution has great potential because it allows reducing the thickness of the packaging without jeopardizing their overall performance at a competitive cost (see scheme in Fig. 8.2).

The light weight packaging concept associated with multifunctional nanocomposite coatings presents environmental and economic benefits such as better waste disposal and reduced energy for production, transport, and storage. Polysaccharide [35], pectin [36], and protein [37] bio-based barrier coatings have been investigated recently with

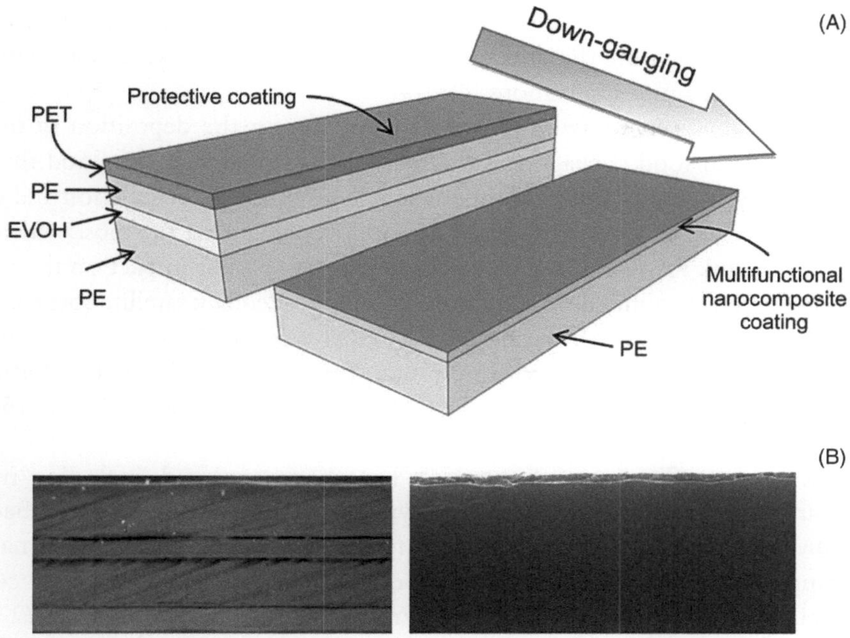

Figure 8.2 Schematic illustration of (A) the light weighting concept associated with multifunctional nanocomposite coatings; (B) cross-section of a multilayer package (total thickness 90 mm) obtained with optical microscopy (left), nanocomposite coating (0.7 mm) on a 12 mm PET substrate (right) obtained with scanning electron microscopy. *(Adapted from I.U. Unalan, G. Cerri, E. Marcuzzo, C.A. Cozzolino, S. Farris, Nanocomposite films and coatings using inorganic nanobuilding blocks (NBB): current applications and future opportunities in the food packaging sector, RSC Adv. 4 (2014) 29393–29428 [6] with permission).*

promising results. However, bio-based barrier nanocoatings are still relatively expensive today compared to conventional synthetic polymeric nanocomposites. Paper/paperboard presents hydrophilic character that limits their potential applications in the food packaging market. However, it has been adapted for food packaging: a nanostructured super hydrophobic surface based on R812S silica nanoparticles and polydimethylsiloxane (PDMS) silicone oil have recently been investigated on paper [38]. A recent study developed antibacterial and barrier paper-based packaging materials based on polystyrene (PS)/TiO_2/Ag nanocomposite coating with promising results [39].

8.2.4 Electrical and electronic properties

In recent years, there has been growing interest in developing smart packaging incorporating electronic devices (e.g., radiofrequency antennas, temperature sensors) into flexible substrates and using cost-effective methods, like printing technologies, to do so. The development of inks with specific rheological properties is at the core of such intelligent packaging. These inks are multicomponent systems containing conductive materials, such as conductors or semiconductors in a liquid vehicle (aqueous or organic), and other additives (i.e., rheology and surface modifiers, binders, humectants, or defoamers) to enable their printing onto the packaging substrate. Electronic inks have recently been investigated for packaging applications [40–44].

A major challenge in applying printing techniques for the deposition of the conductive materials in food packaging is the formulation of suitable inks, and this is the reason why the vast majority of published works focus on their formulation and control [45–50]. In this sense, viscosity, surface tension, and wettability are the most critical characteristics of the ink formulation. These parameters have a clear impact on the printing quality since they determine drop size, drop placement accuracy, satellite formation, and wetting of the substrate. Moreover, the ink should have compatibility with the substrate, good printability, and resolution with minimum printer maintenance. In addition, the ink should be processable (annealing, curing) at temperatures generally below 150°C to be compatible with flexible substrates.

Different types of conductive materials have been explored, including metallic, carbon nanomaterials and conducting polymers. Metal-based inks have been the most studied formulations with a particular focus on inks containing metal nanoparticles, complexes, and organometallic compounds. The best candidates for conductive materials are silver ($\sigma_{Ag} = 6.3 \cdot 10^7\ \Omega^{-1} \cdot m^{-1}$), copper ($\sigma_{Cu} = 5.96 \cdot 10^7\ \Omega^{-1} \cdot m^{-1}$), gold ($\sigma_{Au} = 4.4 \cdot 10^7\ \Omega^{-1} \cdot m^{-1}$) or aluminum ($\sigma_{Al} = 3.8 \cdot 10^7\ \Omega^{-1} \cdot m^{-1}$). At present, most commercial conductive inks are based on silver nanoparticles as silver has the highest electrical conductivity among conductive metals and is resistant to oxidation. However, due to its high cost, great efforts are being dedicated to replace it. Aluminum is considered inappropriate for conductive ink formulations because of its very fast oxidation in air (around 100 ps), which results in loss of electrical conductivity [51]. Oxidation of copper is slower than that of aluminum but it also reduces the electrical conductivity of the printed copper. Thus, one of the main challenges to replace silver with much cheaper metals, such as copper or aluminum, is to avoid their fast oxidation at ambient temperature. Some strategies to avoid their oxidation are based on working in an inert atmosphere; use of hydrocarbon solvents; use of low precursor concentration; and nanoencapsulate particles with nonoxidable metals [52–54], which all imply very sophisticated experimental conditions. A drawback when using nanoparticles is their aggregation and precipitation. Therefore, adding stabilizing agent (i.e., polymer, surfactant) is necessary, but the presence of these components prevents the formation of continuous interconnected metallic nanoparticles, resulting in low conductivity values.

Although organometallic compounds used in inks are mainly based on either silver or copper complexes, a wider range of complexes have been developed using either anionic or coordinating ligands (i.e., aminopropoxides, amidinates, or guanidinate complexes). These complexes must undergo a reduction process to obtain metallic silver or copper and a further thermal treatment to allow particle sintering. The advantages of this strategy are the use of low-cost silver and copper precursors and the possibility of using lower temperatures during the thermal treatment. Impurities from the ligands could result in low conductivity.

Carbon nanomaterials such as carbon nanotubes (CNT) or graphene have recently been explored as conductive inks because of their exceptional electrical, optical, and mechanical properties. The high intrinsic electrical conductivity of individual CNTs or graphene sheets is close to the conductivities of metals [55]. The main drawback of these carbon nanomaterials is the formation of large bundles or ropes that result from Van der Waals forces, causing agglomerates [55,56]. For this reason, dispersant agents must be added to the ink formulation. However, the presence of all these organic additives prevents electrical contact between individual CNTs or graphene sheets which can result in low conductivity films [55,56]. These organic components can be removed by thermal treatment of the printed ink, but again, relatively high temperatures (>200°C) are required for their use which restrict the substrates to which they are printed.

Finally, conducting polymer-based inks have also been developed for printed electronics. Among the different conducting polymers, poly(3,4-ethylendioxythiophene) (PEDOT) has been the most widely studied because of the good conductivity values (300 S/cm) [57]. Despite the low cost, easy conversion into inks, and easy processing, they have shown less conductivity than conventional metals and are also limited in terms of chemical and thermal stability. Individual types of conductive materials have not shown satisfactory properties as ink constituents and as a result the development of nanocomposites has received increased attention.

8.2.5 Optical properties

The main optical properties required in the food packaging sector are light stability and transparency. Light stability is based on preventing color changes (i.e., yellowing) due to photo-oxidation of photosensitive food (e.g., meat, beer, and milk) from UV radiation (wavelength below 340 nm). Light transparency consists of high transmittance in the wavelength of visible light (between 340 and 800 nm). Transparent packaged food can be directly evaluated by the consumer by visual inspection. UV stabilization is generally achieved by using the highly efficient family of hindered amine light stabilizers (HALS) which has capacity to trap radicals. Free radicals are eliminated before they are converted into inert derivatives. There are many formulations based on HALS in the market: a suitable selection of HALS depends on the polymeric matrix (or binders in food packaging coatings) and the target application [58]. HALS additives are combined with ultraviolet light absorbers (UVAs) to prevent color change and delamination of coatings, adhesives, and sealants: the combination of these two chemistries is highly synergistic [59]. However, higher thermal stability than that offered by UVA/HALS systems is often required [60]. In that sense, UV-shielding/absorbing properties based on the use of metal oxides such as TiO_2 [61,62], ZnO [63], or clays [64] have recently been investigated but light transparency is still deficient. More research effort and optimization of the type of nanofiller(s), their associated structure, and processing conditions: time, temperature, energy input, amount of additives, solvents and cosolvents, among others, is needed [6].

8.2.6 Antimicrobial properties

Health risks associated with microbial contamination continues to be one of the main public and governmental concerns concerning food packaging. Despite the evident progress in health risk assessment throughout manufacture, transport, and commercialization of food products, the incidence of foodborne illnesses in developed countries has not waned in the last few decades [65]. In this context, a combination of new technologies with antimicrobial properties as part of food contact products can result in shelf-life extension and foods with improved quality and safety characteristics. In this book, the toxic effect of engineered nanoparticles on microorganisms are discussed in great detail in Chapter 6.

In antimicrobial packaging, a substance with biocide properties is included in the packaging system to extend shelf life and reduce the risk of contamination by pathogens [66]. This aim is achieved by different strategies including the following:

1. The addition of sachets or pads containing volatile antimicrobial agents. This approach is only feasible with volatile compounds, which extremely limits the range of antimicrobials that can be used [67].
2. The incorporation of volatile or nonvolatile antimicrobial agents directly into polymers [68] by extrusion/blending and solution casting. However, these technologies can present serious limitations due to the intrinsic lack of thermal stability of many of such active compounds, which can be lost through degradation and evaporation during the heat transfer involved in the preparation of the packaging material [69].
3. The immobilization of antimicrobials to polymers by ionic interaction or covalent linkages [70]. This is probably the least explored approach for antimicrobial packaging applications, possibly because the most commonly studied antimicrobial agents, such as nisin or essential oils, must migrate from the packaging material to be effective.
4. Coating antimicrobials onto polymer surface. This can be carried out by using polymers that are inherently antimicrobial or incorporating antimicrobial compounds into the coating or ink formulations used in food packaging.

Ideally, packaging materials incorporating antimicrobial agents should encompass a broad spectrum of antimicrobial activity at low concentrations; exert no adverse sensory effects on the product; and comply with current food legislation and toxicity levels. Low production costs are also a must considering the very low mark-ups in the food sector. Antimicrobial coatings can cover all of these requirements and, additionally, they present the advantage of preserving the packaging materials' bulk properties. There is therefore interest in the application of thin coatings, rather than modifying in the bulk packaging

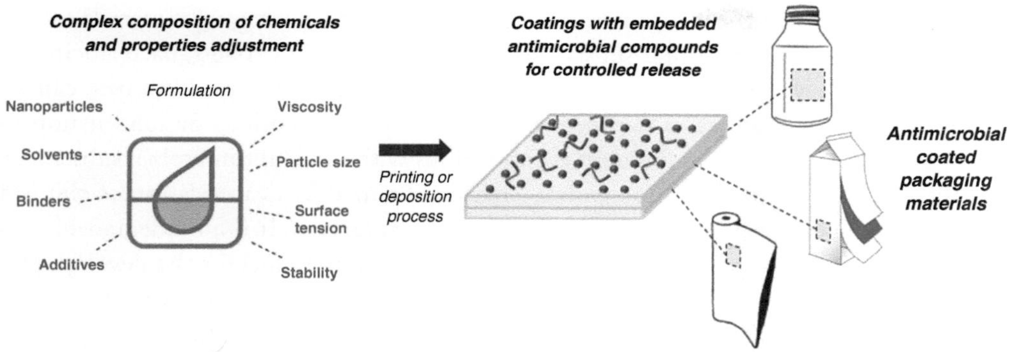

Figure 8.3 ***Factors affecting the composition of antimicrobial ink, and the application of antimicrobial coating in packaging materials.*** *(Adapted from L.J. Bastarrachea, D.E. Wong, M.J. Roman, Z. Lin, J.M. Goddard, Active packaging coatings review, Coatings 5 (2015) 771–791 [73]).*

material, as thin coatings are not expected to affect packaging material's relevant physical and chemical properties [71].

The most important parameter by which to judge the effectiveness of the antimicrobial coatings is the ability of controlled release (see scheme of antimicrobial coatings in Fig. 8.3). Antimicrobial coatings should control the release rates of the incorporated active substances to the surrounding media in a targeted way to achieve the maximum function [72].

There are wide number of active agents with size in the nanometer range that can be incorporated into coatings for controlling growth of pathogenic and/or spoilage microorganisms in packaged products. Metallic nanoparticles and metal oxides are of great interest in this regard due their broad spectrum of action (including Gram positive, Gram negative bacteria, yeasts, and viruses); greater chemical and thermal stability as compared to organic antimicrobials; and involve easy and cost-effective production [74]. Nanosized TiO_2 is one of the most promising antimicrobials that can be incorporated into coatings for diminishing the risks associated with biofilms in food contact surfaces. In relation to their mechanism of action, photocatalytic titanium dioxide nanoparticles induce the production of reactive oxygen species (ROS), glutathione depletion, and cell membrane disruption [75]. Among metallic cations, ionic silver is known to have the greatest antimicrobial capacity against bacteria, yeast, or viruses; it has long-term biocide properties while at the same time being much less toxic to eukaryotic cells and nontoxic to humans [76]. The effects of silver are being reviewed and research on its toxicity is still in progress. Although less studied than silver nanoparticles, copper [77] or zinc–oxide-based [78] nanoparticles can provide an interesting route to develop antimicrobial coatings for packaging materials as well. Nanoclays modified with organic compounds, such as quaternary ammonium-modified montmorillonite, represent another promising alternative for coating paper packaging with antimicrobial properties, as reported by Soares et al. [79].

On the other hand, the antimicrobial properties of food packaging materials can be improved by the use of natural substances, such as enzymatic extracts, bacteriocins, or essential oils. They are generally recognized as safe (GRAS) food preservatives, can be incorporated into coatings in the form of polymer binder solutions or solid nanoparticles, and are usually used in different combinations with other antimicrobial substances. For instance, chitosan is an amino-polysaccharide obtained by deacetylation of chitin: a structural component present in the shell of some crustaceans. It is a biodegradable inherently antimicrobial polymer, and it presents promising potential for the development of new antimicrobial packaging concepts. There is extensive literature as to its antimicrobial efficacy either alone or in blends with other polymers [80–82].

Another type of natural antibacterial agent with important industrial applications are bacteriocins. These are peptides produced by a broad range of bacteria, although the great majority of these compounds are ascribed to the genus *Lactococcus*. Nisin is the main bactericin used industrially, and it is for instance applied in the prevention of contamination of cheese surfaces by *Clostridium* species and *L. monocytogenes*. The incorporation of bacteriocins in coatings, such as nisin/methylcellulose [83] or enterocin/PE-PEG block copolymer-based coatings [84] have been investigated for a long time and there is extensive data on their performance. Among enzymes, lysozyme, most commonly extracted from egg white, is capable of breaking the glucosidic bonds in the peptidoglycan of Gram positive bacteria and has been successfully incorporated in zein nanoparticles [85] or cellulose acetate films [86]. Essential oils are also of great importance as natural antimicrobial agents. Some examples include the use of oregano and clove essential oils in methylcellulose films [87] or the addition of cinnamon essential oil to chitosan-based films [88]. However, one of the main drawbacks of incorporating these natural oils into polymers is their high chemical and thermal instability, as well as a very high volatility. Additionally, their effectiveness is much lower than other traditional antimicrobials, which implies the need for higher filler contents. These issues pose difficult challenges as to the development of cost-effective polymer technologies releasing essential oils.

With the range of strategies reviewed, it can be inferred that while the incorporation of active agents and use of specialized packaging processes will increase material costs, there are opportunities for new products with enhanced food safety, and with prospective reduction of waste of packaged goods. Therefore, there is potential for increasing product value through the use of packaging coatings with smart integration of antimicrobial agents.

8.3 COATING/PRINTING TECHNOLOGIES USED IN THE FOOD PACKAGING INDUSTRY

8.3.1 Main printing technologies

There is a wide repertoire of printing methods for many industrial applications [89]: printed electronics [90], magazines, wallpaper, floor covering, and food packaging [91], to name a few. These printing methods include screen printing [92], flexographic [93], offset printing [94], ink-jet printing [95], and gravure printing [94]. The selection of a printing method depends on the ink properties (such as viscosity, surface tension, Fig. 8.3), pattern geometry, speed of the printing process, yield, quality and production costs. Some of the most important printing parameters is listed as follows. These also apply for inks and coatings incorporating.

1. Printing accuracy and resolution.
2. Uniformity when printing from a few squared centimeters to a one squared meter. It will be affected by the ink composition and the drying process.
3. Wetting control and interface formation. The surface energy of the substrate and bonding between the printed layers and the substrate must allow creating a quality printed pattern.
4. The ink compatibility with the printer components, such as rollers, doctor blades, ink-jet heads, has an important impact on the yield and quality when printing in mass production.
5. Process cost considerations.

8.3.2 Flexography

Flexography is the most important process used for printing food flexible packaging. It is also used to print paperbacks, comics, and newspapers. Flexography employs a flexible rubber plate, rather than a metal plate (letterpress), to print onto plastic, corrugated cardboard, and paper-based substrates. This printing process (shown in Fig. 8.4) is based on the transfer of the ink from the ink tray through a fountain cylinder to an Anilox cylinder. Such a cylinder presents texture with small wells holding ink that is transferred to a plate cylinder (flexible rubber-like material). This results in evenly distributed ink with uniform thickness. In the process of transferring ink from the Anilox cylinder to the plate cylinder, there is a blade or scraper, which removes the excess ink onto a flexible plate. To transfer the image, a substrate (food packaging) is sandwiched between the plate cylinder and the impression cylinder. Finally the substrate is then fed through either a dryer drum to dry the ink on the substrate or a UV ray device to cure the final ink (Fig. 8.4).

Flexography inks can be either synthetic or natural resins dissolved in low boiling point solvents, which result in liquid inks with low viscosity, rather than pastes. Common solvents used in these inks are low molecular weight alcohol, such as ethanol, n-propyl, or butanol, mixed with esters and glycol ethers and/or aliphatic hydrocarbons. Polymeric resins in these inks are typically based on nitrocellulose, polyamide, and polyurethane, and the dyes in them are organic molecules or metal complexes. Besides, flexography inks contain plasticizers to provide higher flexibility to printed films and waxes, which add rub resistance. The formulation of the ink, with high solubility of all its components, and high drying speed are desirable for high quality printed images onto the substrate.

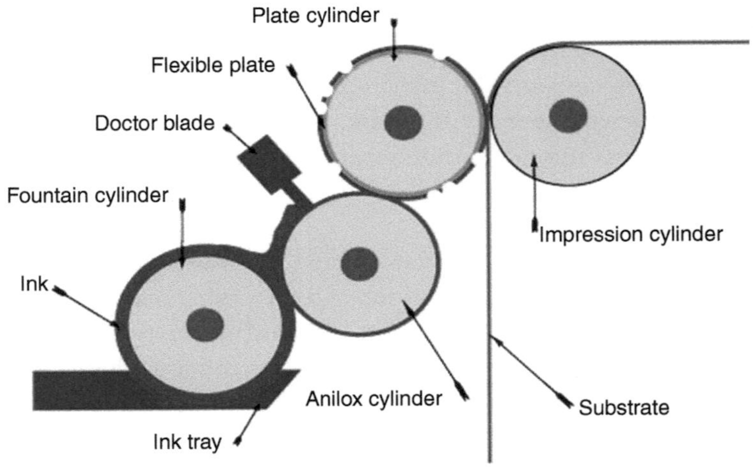

Figure 8.4 ***Flexography printing process.*** *Reproduced from American Label Group Ltd, Explore the flexographic printing process, Available from: http://www.americanlabel.com/algweb/alg2/explore_flexo.asp [96] with permission.*

The main advances in flexographic printing have taken place in improvements in the material of photopolymer printing plates and methods of plate creation [96]. Furthermore, flexographic printing has improved its tonal value (or dot gain) which translates into higher quality and accuracy of printed substrates.

8.3.3 Rotogravure

Rotogravure is another technique applied in printing flexible food packaging. This technology is based on an intanglio process, where a sunken surface is obtained after printing cylinders engrave the desired pattern [97]. In this printing process, a group of micrometric cells (i.e., small cavities engraved onto cylinders) act as ink reservoirs that transfer ink to the surface of packaging substrates by direct contact. The amount of ink contained in the cells translates into different color intensities onto the substrate, and, therefore, the cell dimensions must be designed according to the intensity of the ink color desired onto the substrate. Thus, the deeper the cells, the brighter the colors will be. A rotogravure printing press has a printing unit for each color, hence, the number of units will depend on the color required in the final image.

The printing process with rotogravure is shown in Fig. 8.5. A blade scrapes off the excess ink around the noncontact printing areas of the engraved cylinder, leaving the amount of ink required in the engraved cells ready to be transferred to the substrate. An impression roller is used to apply force and press the substrate onto the engraved cylinder to transfer the ink and ensure an even ink cover. Finally, the printed substrate passes through a dryer unit before moving to the next color unit for the next application of ink.

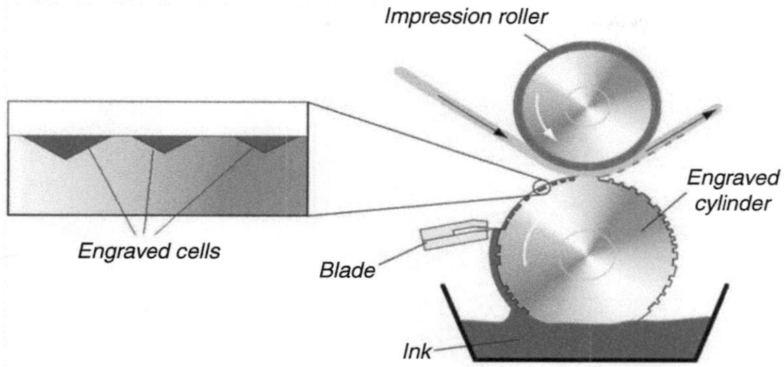

Figure 8.5 ***Rotogravure printing process.*** *Reproduced from Discovery flexibles, Rotogravure process, Available from: http://www.discoveryflexibles.com/rotogravure/ [98] with permission.*

Gravure inks are very similar to the flexographic inks used in food packaging and are constituted by organic solvents such as esters and alcohols. These present low boiling point, and therefore become volatilized during the drying step, which minimizes their migration from the printed packaging to foodstuff. A number of packaging products include resins (i.e., cellulose nitrate, maleic, acrylates, polyurethanes, polyamides); plasticizers (i.e., phthalates, citrates, adipates), dyes, and pigments. Besides organic based inks, water-based gravure inks are also widely used in the food packaging sector.

8.3.4 Ink-jet

Ink-jet printing is a noncontact dot matrix printing technology in which ink droplets are jetted from small nozzles directly to a specified position on a media (substrate) to create an image [90,91,95]. The principle where a liquid stream can be broken into small droplets with uniform size and spacing was introduced by Lord Rayleigh in 1878. Based on the mechanism of droplet formation through the breakup of a continuous liquid jet, two main ink-jet technologies were designed [89]. The formation of drops by thermal and piezoelectric mechanisms is at the core of the main ink-jet technology. A second main technology, continuous ink-jet, involves the uninterrupted generation of drops from a drop generator. This breaks off the ink in small electrically charged droplets. Drops then go through an electrified plate which deflects the noncharged droplets to the gutter, where they will be recirculated, whereas the charged droplets will be directed to the substrate.

Ink-jet printing has become the main print technology in display applications. Special interest has been focused on UV-continuous ink-jet for printing polymeric food packaging such as polyethylene and polypropylene. This UV-continuous ink-jet printing technique is growing very fast due to its good jetting performance and robust printing. It provides high image quality and durability (i.e., suitable adhesion, scratch

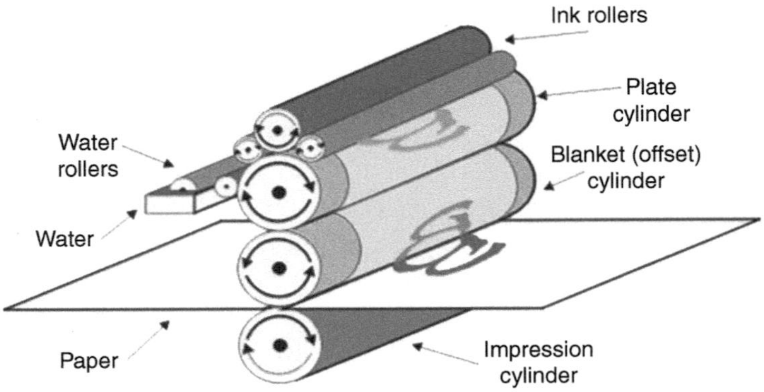

Figure 8.6 ***Offset printing process.*** *Reproduced from Offset printing technology, offset lithography, 2016. Available from: http://www.offsetprintingtechnology.com/sub-categories/offset-lithography/ [99] with permission.*

resistance, and solvent resistance), high curing speed, and curing degree (92–98%), which minimizes the migration of unreacted monomers from the curing process to the foodstuff, and provides good optical properties to the printed media, such as gloss, color, or brightness.

8.3.5 Off-set

In the offset printing process, the surface with and without printed image lies on the same plane and differ in their wetting properties. The nonimage area is hydrophilic and therefore can be wetted with water and a mixture of both water and alcohol. In contrast, the image area (ink-receptive areas) is hydrophobic and can be wetted with oil. The off-set printing process is shown in Fig. 8.6.

The ink used in off-set printing must present high viscosity, and its temperature should be between 25 and 40°C to avoid the evaporation of the solvent. The resins used in the ink need to be soluble in the solvent, stabilize the ink in the inking system and ensure its good drying. The solvents used for the offset printing process are normally vegetable oils (which present a few double chemical bonds), mineral oils, and fatty acid esters. The latter ones are used in food packaging printing due to their low level of migration from the printed polymeric material to the foodstuff [94].

8.3.6 Other technologies

There is an increasing trend to use new packaging materials but improvements in the sealability of such laminated composite materials are required [100]. This is carried out with both new and conventional approaches. Among the new technologies, nanoimprinted lithography and injection mold coatings are especially relevant [101]; and also extrusion and lamination coatings technologies, which are conventionally used in the food packaging sector. Extrusion coating consists of heating together a mixture of two different polymer granules or pellets. The combined polymers are extruded (pushed through a die of a specific cross-section) as a single product which is a multilayer web of plastic film. Cast extrusion and blown extrusion are the two extrusion methods usually considered [91,98]. The lamination processing consists of two or more polymers in web form which will become multiple polymeric layers with enhanced properties. This is achieved by extrusion or hot melt lamination [91,102]. A wide variety of printing technologies are in use for printing coatings and inks containing nanomaterials and their effectivity will have important nanosafety implications.

8.4 NANOSAFETY AND NANOTOXICOLOGY

Specific regulations for nanomaterial-based products are into debate worldwide due to their growing market and nanosafety concerns. In this book, Chapter 9 exposes how the regulatory bodies are reacting to ensure food safety and the existing principles applied to nanofoods; however contact materials have not been dealt specifically in Chapter 9.

The already established Regulation (EC) 1935/2004 for food contact materials, and a revised Regulation 10/2011 for "Plastic Food Contact Materials" address nanomaterials nanosafety and nanotoxicology [103]. The general requirements of Regulation (EC) 1935/2004 are laid down in Article 3, where it is stated that materials coming into contact with food shall be manufactured such that, under normal and foreseeable conditions of use, they would not transfer their constituents to food in quantities which could endanger human health or bring about an unacceptable change in the composition of food or deterioration in the organoleptic characteristics thereof. In Article 5, specific measures for groups of materials and articles are defined, but no positive list of allowed substances with specific migration limits for printing inks exists. Therefore the compliance of nanobased coatings and inks can only be assessed on the basis of the overall migration limits. Regarding the risks of substances in adhesives, coatings, and inks, the European Commission has planned to launch a consultation on policy options.

Hannon et al. [104] reviewed the main challenges related to the use of engineered nanoparticles in food contact materials. This work shows that there are substantial gaps of knowledge regarding the use of nanobased coatings/inks in food packaging with reference to human health risk assessment. Migration phenomena are influenced by temperature, time, concentration gradient, material properties, migrant material position in the material, the interaction between the migrating material, and the food product and physico-chemical conditions in the food product such as pH. The presence of nanomaterials in coatings and inks for food packaging materials can cause migration of nanomaterials to the outermost layers of the food products. More research is needed on the migration of nanoparticles from food contact materials, their diffusion mechanisms and their associated nanosafety implications for consumer health.

8.5 FUTURE TRENDS AND CHALLENGES

The main challenges of nanocoatings and inks for commercial food packaging are related to the improvement of their functionality at reasonable costs. Additionally, progress in the assessment of their nanosafety is needed. These challenges are in agreement with the global challenges of the packaging industry, some of the most important being the need to: (1) increase sustainability of manufacturing processes, (2) improve recyclability of materials, and (3) improve performance and functionality. With reference to (3) in particular, there is a need to overcome adhesion issues and achieve higher barrier properties. Adhesion problems in laminates and nanobased functional inks are solved currently by the development and application of specific surface treatments such as those based on atmospheric pressure plasma technology, but greater research effort is needed to improve nanocoating and nanoink formulations without compromising their functionality.

Keeping or improving barrier properties of packaging to extend food products' shelf life with less amount of thermoplastic commodity is still a challenge [105]. Greater gas and water vapor barrier properties are required. In particular, two well-accepted needs are materials with greater oxygen barrier for packaging transparent to visible light. This type of packaging is of great importance for fresh food (i.e., vegetables, fruits); for lidding films for retort products; and low cost water repellent coatings for frozen food (i.e., meat, fish) such as packaging based on cardboard packaging. Nanobased barrier coatings are a promising alternative for overcoming these challenges.

Multifunctional nanobased coating and inks conferring improved mechanical, barrier, electrical, optical, and antimicrobial properties are in great demand by the packaging industry. The development of intelligent packaging is advancing toward improving the protection and communication aspects of packaging materials [105,106]. Active packaging (packaging which interacts with the packaged food) requires to extend food shelf life with antimicrobials, and also by using oxygen [107] and ethylene [108] scavengers, based mostly on titanium dioxide polymeric nanocomposite coatings. Intelligent packaging (which senses and communicates information from the packaged product) involves using functional nanomaterial-based ink formulations with innovative electronic designs and architectures printed on packaging materials to achieve several functionalities, such as communication by radiofrequency, time-temperature indicators, gas sensors, and freshness indicators, among others. New concepts such as integrated printed radiofrequency antennas; printed temperature sensors; and printed supercaps, through the use of electronic inks, are still under development [109]. A promising concept developed for oxygen sensors based on photocatalytic titanium dioxide-nanoparticle-based inks consists of reversible multistep redox consecutive reactions using UV radiation, titanium dioxide nanoparticles, a sacrificial electron donor (which can be glycerol), and an encapsulated dye (i.e., methylene blue dye) [110]. Other trends with great impact in intelligent packaging are focused on the detection of spoilage and pathogenic microorganisms [105]. They are based on the use of quantum dots, and fluorescent organic compounds and nanoparticles, which offer high sensitivity in the detection. A successful example was the selective detection of two different species from *Escherichia coli* and *Salmonella typhimurium* with quantum dot–antibody conjugates [111]. This approach opened the door to detecting multiple species of bacteria. Fluorescent nanoparticles are also applied as highly sensitive sensors for proteins, enzymes, DNA, bacteria, and chemicals related to changes in food such as CO_2 or ammonia [112]. The application of nanobased inks to detect biological indicators of food spoilage is expected to grow considerably in the near future. Time, temperature, humidity, and freshness indicators can already be monitored by using nanobased coatings and inks. The presence of intelligent packaging solutions in the market such as those briefly exposed is expected to increase in the next few years.

As a final point, there is a global trend to extend the variety of food products offered to the consumer with short delivery dates. Regarding coating/printing technologies, this trend involves a progressive move toward using highly versatile environment friendly ink-jet technology, which is also prepared for fast customization, and replace gravure and off-set and flexography technologies. The trend in nanobased ink formulations, therefore, is to design and develop functional, sustainable, and reliable inks and coatings inks at lower costs.

8.6 CONCLUSIONS

Coating and ink formulations including nanomaterials are increasing their application in food packaging. Some nanobased products with enhanced abrasion and scratch resistance, oxygen and water vapor barrier, and UV absorption properties have already been developed with demonstrated effectivity. In nanobased coatings and ink formulations, there could be the solution to overcome some main challenges in the food packaging sector such as improving the shelf life of raw materials and recyclability, which will in turn decrease waste, and play a key part in active and intelligent packaging. Some of the most interesting examples that are under investigation are novel coatings to achieve transparent high barrier properties; inks with enhanced electrical conductivity using room temperature sintering; and effective, durable, and safe antimicrobial varnishes for meat packaging, among others. Innovative coating and printing technologies required for nanobased inks could be complementary to conventional off-set, flexography, rotogravure, screen, or pad printing technologies. Roll-to-roll nanoimprinted lithography is a potential forthcoming technology in the food packaging sector.

The growing market of nanobased inks and coatings increases nanosafety concerns and there is urgent need to accelerate the development of specific standards to validate the nanosafety and nanotoxicology of the developed food contact products containing nanomaterials. It can be concluded that the design and development of functional, sustainable, and safe nanobased inks and coatings for food packaging with a suitable performance at acceptable costs presents satisfactory and very promising perspectives in the near future.

REFERENCES

[1] D. Dainelli, N. Gontard, D. Spyropoulos, E. Zondervan-van den Beuken, P. Tobback, Active and intelligent food packaging: legal aspects and safety concerns, Trends Food Sci. Technol. 19 (2008) S103–S112.
[2] J.M. Fuentes-Alventosa, L. Introzzi, N. Santo, G. Cerri, A. Brundu, S. Farris, Self-assembled nanostructured biohybrid coatings by an integrated 'sol–gel/intercalation' approach, RSC Adv. 3 (2013) 25086–25096.
[3] N.A. El-Wakila, E.A. Hassana, R.E. Abou-Zeida, A. Dufresne, Development of wheat gluten/nanocellulose/titanium dioxide nanocomposites for active food packaging, Carbohydr. Polym. 124 (2015) 337–346.
[4] H.K. Kim, S.J. Kim, H.S. Lee, J.H. Choi, C.M. Jeong, M.H. Sung, S.-H. Park, H.J. Park, Mechanical and barrier properties of poly(lactic acids) films coated by nanoclay-ink composition, J. Appl. Polym. Sci. 127 (5) (2013) 3823–3829.
[5] Z. Shariatiania, M. Fazli, Mechanical properties and antibacterial activities of novel nanocomposite films of chitosan and starch, Food Hydrocolloids 45 (2015) 112–124.
[6] I.U. Unalan, G. Cerri, E. Marcuzzo, C.A. Cozzolino, S. Farris, Nanocomposite films and coatings using inorganic nanobuilding blocks (NBB): current applications and future opportunities in the food packaging sector, RSC Adv. 4 (2014) 29393–29428.
[7] X.H. Li, Y.G. Xing, W.L. Li, Y.H. Jiang, Y.L. Ding, Antibacterial and physical properties of poly(vinyl chloride)-based film coated with ZnO nanoparticles, Food Sci. Technol. Int. 16 (3) (2010) 225–232.
[8] F. Li, P. Biagioni, M. Bollani, A. Maccagnan, L. Piergiovanni, Multi-functional coating of cellulose nanocrystals for flexible packaging applications, Cellulose 20 (2013) 2491–2504.
[9] R.H. McMullin, New aspects in the sustainability of nanoparticle-modified, in: 38th Annual Waterborne Symposium, New Orleans, 2011. Available from: http://www.byk.com/fileadmin/byk/articles_lectures/Nano_Sustainability.pdf
[10] S.S. Ray, M. Okamoto, Polymer/layered silicate nanocomposites: a review from preparation to processing, Prog. Polym. Sci. 28 (11) (2003) 1539–1641.
[11] C. Silvestre, D. Duraccio, S. Cimmino, Food packaging based on polymer nanomaterials, Prog. Polym. Sci. 36 (2011) 1766–1782.
[12] J. Schneider, M.I. Akbar, J. Dutroncy, D. Kiesler, M. Leins, A. Schulz, M. Walker, U. Schumacher, U. Stroth, Silicon oxide barrier coatings deposited on polymer materials for applications in food packaging industry, Plasma Processes Polym. 6 (1) (2009) S700–S704.
[13] Amcor, Ceramis, 2016. Available from: http://www.amcor.com
[14] R. Sang, Thin film encapsulation for OLED display using silicon nitride and silicon oxide composite film, in: International Conference on Electronic Packaging Technology & High Density Packaging, 2011, pp. 1175–1178.
[15] J. Garcia-Torres, D. Sylla, L. Molina, E. Crespo, J. Mota, L. Bautista, Surface modification of cellulosic substrates via atmospheric pressure plasma polymerization of acrylic acid: structure and properties, Appl. Surf. Sci. 305 (2014) 292–300.
[16] X.Z. Tang, P. Kumar, S. Alavi, K.P. Sandeep, Recent advances in biopolymers and biopolymer-based nanocomposites for food packaging materials, Crit. Rev. Food Sci. Nutr. 52 (2012) 426–442.
[17] N. Borse, T. Hutley, Polymer/clay nanocomposites: an emerging material class, 4th Saudi Tech. Conf. & Exhib., Riyadh, III, 2006, pp. 237-243.
[18] P.M. Ajayan, L.S. Schadler, P.V. Braun, Nanocomposite Science and Technology, Wiley-VCH GmBH & Co. KGaA, Weinheim, (2003).

[19] S.-H. Park, S.J. Kim, H.S. Lee, J.H. Choi, C.M. Jeong, M.H. Sung, D.H. Kim, H.J. Park, Improvement of oxygen barrier of oriented polypropylene films coated by gravure ink-containing nanoclays, J. Appl. Polym. Sci. 121 (3) (2011) 1788–1795.
[20] N. Bumbudsanpharoke, S. Ko, Nano-food packaging: an overview of market, migration research, and safety regulations, J. Food Sci. 80 (5) (2015) R910–R923.
[21] N.B. Hatzigrigoriou, C.D. Papaspyrides, Nanotechnology in plastic food-contact materials, J. Appl. Polym. Sci. 122 (6) (2011) 3719–3738.
[22] Nanocor®, Film and sheet applications, technical bulletin, 1-7. Available from: http://www.nanocor.com/tech_sheets/I105.pdf
[23] InMat Inc., Ultrahigh barrier coating breakthrough from InMat® reduces food packaging costs and environmental impact, 2009. Available from: http://www.inmat.com/upload/files/inmat_press_release_05072009.pdf
[24] NanoPack, NanoSeal™—Barrier coating, OPP barrier coated film, polypropylene film, liquid coating, and 2016. Available from: http://www.nanopackinc.com/products.asp
[25] G. Qian, D. Jarus, V. Zabrocki, Use of organoclay in HDPE nanocomposites to provide barrier properties in containers and film (2009) US Patent US7,629,406B2.
[26] K. Khemani, G. Moad, E. Lascaris, G. Li, G. Simon, J. Habsuda, R. Shanks, A. Genovese, W. Gunaratne, L. Nichols, Starch nanocomposite materials (2008) US Patent US20,100,307,951A1.
[27] K. Tadanaga, K. Iwashita, T. Minami, Coating and water permeation properties of SiO_2 thin films prepared by the sol-gel method on nylon-6 substrates, J. Sol-Gel Sci. Technol. 6 (1) (1996) 107–111.
[28] S. Amberg-Schwab, H. Katschorek, U. Weber, M. Hoffmann, A. Burger, Barrier properties of inorganic-organic polymers: influence of starting compounds, curing conditions and storage-scaling-up to industrial application, Sol-Gel Sci. Technol. 19 (1) (2000) 125–129.
[29] S. Farris, L. Introzzi, J.M. Fuentes-Alventosa, N. Santo, R. Rocca, L. Piergiovanni, Self-assembled pullulan–silica oxygen barrier hybrid coatings for food packaging applications, J. Agric. Food Chem. 60 (2012) 782–790.
[30] K. Tadanaga, K. Iwashita, T. Minami, N. Tohge, Coating and water permeation properties of SiO_2 thin films prepared by the sol-gel method on nylon-6 substrates, J Sol-Gel Sci. Technol. 6 (1) (1996) 107–111.
[31] K. Jang, H. Kim, The gas barrier coating of 3-aminopropyltriethoxysilane on polypropylene film, J. Sol-Gel Sci. Technol. 41 (1) (2007) 19–24.
[32] M. Iotti, P. Fabbri, M. Messori, F. Pilati, P. Fava, Organic-inorganic hybrid coatings for the modification of barrier properties of poly(lactic acid) films for food packaging applications, J. Polym. Environ. 17 (1) (2009) 10–19.
[33] G. Bang, S.W. Kim, Biodegradable poly (lactic acid)-based hybrid coating materials for food packaging films with gas barrier properties, J. Ind. Eng. Chem. 18 (2012) 1063–1068.
[34] S. Kwak, J. Jun, E.-S. Jung, Gas barrier film with a compositional gradient interface prepared by plasma modification of an organic/inorganic hybrid sol–gel coat, Langmuir 25 (14) (2009) 8051–8055.
[35] S. Farris, L. Introzzi, J.M. Fuentes-Alventosa, N. Santo, R. Rocca, L. Piergiovanni, Self-assembled pullulan–silica oxygen barrier hybrid coatings for food packaging applications, J. Agric. Food Chem. 60 (3) (2012) 782–790.
[36] V.D. Alves, Barrier properties of biodegradable composite films based on kappa-carrageenan/pectin blends and mica flakes, Carbohydr. Polym. 79 (2010) 269–276.
[37] F. Tihminlioglua, İ.D. Atika, B. Özenb, Water vapor and oxygen-barrier performance of corn–zein coated polypropylene films, J. Food Eng. 96 (2010) 342–347.
[38] W. Chen, X. Wang, Q. Tao, J. Wang, Z. Zheng, X. Wang, Lotus-like paper/paperboard packaging prepared with nanomodified overprint varnish, Appl. Surf. Sci. 266 (2013) 319–325.
[39] A.M. Youssef, S. Kamel, M.A. El-Samahy, Morphological and antibacterial properties of modified paper by PS nanocomposites for packaging applications, Carbohydr. Polym. 98 (1) (2013) 1166–1172.
[40] T. Bakhishev, V. Subramanian, Investigation of gold nanoparticle inks for low-temperature lead-free packaging technology, J. Electron. Mater. 38 (12) (2009) 2720–2725.

[41] J. Felba, K. Nitsch, T Piasecki, S Tesarski, A Moscicki, A. Kinart, D. Bonfert, K. Bock, Properties of conductive microstructures containing nano sized silver particles, in: Proc. 2009 11th IEEE Electronics Packaging Technology Conference, 2009, pp. 879–883.
[42] J. Szczech, C.M. Megaridis, D.R. Gamota, J. Zhang, Fine-line conductor manufacturing using drop-on-demand PZT printing technology, IEEE Trans. Electron. Packag. Manuf. 25 (1) (2002) 26–33.
[43] D.J. Hayes, W.R. Cox, D. B. Wallace, Printing system for MEMS packaging, in: Proc. 2001 SPIE Micromachining & Microfabrication Conference, 2001, pp. 22–25.
[44] O. Gumeniuk, V. Sherstyuk, Nanotechnologies in printing and packaging, Sci. Technol. (2009) 32–41.
[45] S. Magdassi, Ink requirements and formulation guidelines, in: S. Magdassi (Ed.), The Chemistry of Ink-Jet Inks, World Scientific, NJ, London, Singapore, 2010, pp. 19–41.
[46] J.A. Lewis, Direct ink writing of 3D functional materials, Adv. Funct. Mater. 16 (2006) 2193–2204.
[47] G.M. Lous, I.A. Cornejo, T.F. McNulty, A. Safari, S.C. Danforth, Fabrication of piezoelectric ceramic/polymer composite transducers using fused deposition of ceramics, J. Am. Chem. Soc. 83 (2000) 124–128.
[48] A. Kyamyshny, J. Steinke, S. Magdassi, Metal-based inkjet inks for printed electronics, Open Appl. Phys. J. 4 (2011) 19–36.
[49] B.-J. de Gans, P.C. Duineveld, U.S. Schubert, Inkjet printing of polymers: state of the art and future developments, Adv. Mater. 16 (3) (2004) 203–213.
[50] J.A. Lewis, J.E. Smay, J. Stuecker, J. Cesarano III, Direct ink writing of three-dimensional ceramic structures, J. Am. Ceram. Soc. 89 (12) (2006) 3599–3609.
[51] T. Campbell, R.K. Kalia, A. Nakano, P. Vashishta, S. Ogata, S. Rodgers, Dynamics of oxidation of aluminum nanoclusters using variable charge molecular-dynamics simulations on parallel computers, Phys. Rev. Lett. 82 (1999) 4866–4869.
[52] M. Grouchko, A. Kamyshny, S. Magdassi, Formation of air-stable copper–silver core–shell nanoparticles for inkjet printing, J. Mater. Chem. 19 (2009) 3057–3062; F. Douglis, Th. Ball, Tracking and viewing changes on the web, in: Proc. 1996 USENIX Technical Conference, 1996.
[53] S. Magdassi, M. Grouchko, A. Kamyshny, Conductive ink-jet inks for plastic electronics: air stable copper nanoparticles and room temperature sintering, in: Proc. 2009 NIP25 & Digital Fabrication. Technologies Program Proc, 2009, pp. 611–613.
[54] S. Magdassi, M. Grouchko, A. Kamyshny, Copper nanoparticles for printed electronics: routes towards achieving oxidation stability, Materials 3 (9) (2010) 4626–4638.
[55] L. Hu, D.S. Hecht, G. Grüner, Carbon nanotube thin films: fabrication, properties, and applications, Chem. Rev. 110 (2010) 5790–5844.
[56] A. Thess, R. Lee, P. Nikolaev, H. Dai, P. Petit, J. Robert, C. Xu, Y.H. Lee, S.G. Kim, A.G. Rinzler, D.T. Colbert, G.E. Scuseria, D. Tomanek, J.E. Fischer, R.E. Smalley, Crystalline ropes of metallic carbon nanotubes, Science 273 (5274) (1996) 483–487.
[57] M.-C. Choi, Y. Kim, C.-S. Ha, Polymers for flexible displays: from material selection to device applications, Prog. Polym. Sci. 33 (6) (2008) 581–630.
[58] K.J. Valentas, E. Rotstein, R. Paul Singh (Eds.), Handbook of Food Engineering Practice, CRC Press, Boca Raton, New York, 1997, pp. 736.
[59] BASF, Tinuvin® high performance light stabilizers. Available from: http://product-finder.basf.com/group/corporate/product-finder/en/brand/TINUVIN
[60] R.H.D. Beswick, D.J. Dunn, Plastics in Packaging, A Rapra Market Report, Rapra Technology Ltd, Shawbury, Shrewsbury, Shropshiore, UK, (2002).
[61] B. Seentrakoon, B. Junhasavasdikul, W. Chavasiri, Enhanced UV-protection and antibacterial properties of natural rubber/rutile-TiO_2 nanocomposites, Polym. Degrad. Stab. 98 (2013) 566–578.
[62] S.-H. Hsieh, F.-R. Zhang, H.-S. Li, Anti-ultraviolet and physical properties of woolen fabrics cured with citric acid and TiO_2/chitosan, J. Appl. Polym. Sci. 100 (6) (2006) 4311–4319.
[63] Q.G. Du, G. Alagappan, H. Dai, H.V. Demir, H.Y. Yu, X.W. Sun, C.H. Kam, UV-blocking ZnO nanostructure anti-reflective coatings, Opt. Commun. 285 (2012) 3238–3241.
[64] J. Vartiainen, M. Tuominen, K. Nattinen, Bio-hybrid nanocomposite coatings from sonicated chitosan and nanoclay, J. Appl. Polym. Sci. 116 (2010) 3638–3647.

[65] A.H. Havelaar, S. Brul, A. de Jong, R. de Jonge, M.H. Zwietering, B.H. ter Kuile, Future challenges to microbial food safety, Int. J. Food Microbiol. 139 (1) (2010) S79–S94.
[66] S. Quintavalla, L. Vicini, Antimicrobial food packaging in meat industry, Meat Sci. 62 (3) (2002) 373–380.
[67] T. Padgett, I.Y. Han, P.L. Dawson, Incorporation of food-grade antimicrobial compounds into biodegradable packaging films, J. Food Prot. 61 (10) (1998) 1330–1335.
[68] J.H. Han, J.D. Floros, Casting antimicrobial packaging films and measuring their physical properties and antimicrobial activity, J. Plast. Film Sheet. 13 (1997) 287–298.
[69] P. Suppakul, J. Miltz, K. Sonneveld, S.W. Bigger, Active packaging technologies with an emphasis on antimicrobial packaging and its applications, J. Food Sci. 68 (2003) 408–420.
[70] F. Kong, Y.F. Hu, Biomolecule immobilization techniques for bioactive paper fabrication, Anal. Bioanal. Chem. 403 (2012) 7–13.
[71] B.D. Ratner, Surface modification of polymers: chemical, biological and surface analytical challenges, Biosens. Bioelectron. 10 (9-10) (1995) 797–804.
[72] K. Yam, P. Takhistov, J. Miltz, Intelligent packaging: concepts and applications, J. Food Sci. 70 (2005) R1–R10.
[73] L.J. Bastarrachea, D.E. Wong, M.J. Roman, Z. Lin, J.M. Goddard, Active packaging coatings review, Coatings 5 (2015) 771–791.
[74] H. Plaza, Antimicrobial polymers with metal nanoparticles, Int. J. Mol. Sci. 16 (2015) 2099–2116.
[75] A.E. Nel, L. Mädler, D. Velegol, T. Xia, E.M. Hoek, P. Somasundaran, F. Klaessig, V. Castranova, M. Thompson, Understanding biophysicochemical interactions at the nanobio interface, Nat. Mater. 8 (7) (2009) 543–557.
[76] M.R. de Moura, L.H.C. Mattoso, V. Zucolotto, Development of cellulose-based bactericidal nanocomposites containing silver nanoparticles and their use as active food packaging, J. Food Eng. 109 (3) (2012) 520–524.
[77] A. Llorens, E. Lloret, P.A. Picouet, R. Trbojevich, A. Fernandez, Metallic-based micro and nanocomposites in food contact materials and active food packaging, Trends Food Sci. Technol. 24 (1) (2012) 19–29.
[78] P.J.P. Espitia, N. Ferreira Soares, J.S. dos Reis Coimbra, N.J. de Andrade, R. Souza Cruz, E.A. Alves Medeiros, Zinc oxide nanoparticles: synthesis, antimicrobial activity and food packaging applications, Food Bioprocess Technol. 5 (5) (2012) 1447–1464.
[79] N.F.F. Soares, F.K.V. Moreira, T.L. Fialho, N.R. Melo, Triclosan-based antibacterial paper reinforced with nano montmorillonite: a model nanocomposite for the development of new active packaging, Polym. Adv. Technol. 23 (5) (2012) 901–908.
[80] P.K. Dutta, S. Tripathi, G.K. Mehrotra, J. Dutta, Perspectives for chitosan based antimicrobial films in food applications, Food Chem. 114 (4) (2009) 1173–1182.
[81] M. Aider, Chitosan application for active bio-based films production and potential in the food industry: review, Food Sci. Technol. 43 (6) (2010) 837–842.
[82] P. Fernandez-Saiz, M.J. Ocio, J.M. Lagaron, Antibacterial chitosan-based blends with ethylene-vinyl alcohol copolymer, Carbohydr. Polym. 80 (3) (2010) 874–884.
[83] P. Appendini, J.H. Hotchkiss, Immobilization of lysozyme on synthetic polymers for the application of food packages, Annual Meeting of Institute of Food Technologists, New Orleans, 1996, Book of Abstract, p. 177.
[84] R. Iseppi, F. Pilati, M. Marini, M. Toselli, S. de Niederhäusern, E. Guerrieri, P. Messi, C. Sabia, G. Manicardi, I. Anacarso, M. Bondi, Anti-listerial activity of a polymeric film coated with hybrid coatings doped with Enterocin 416K1 for use as bioactive food packaging, Int. J. Food Microbiol. 123 (3) (2008) 281–287.
[85] Ç. Mecitoğlub, A. Yemenicioğlua, A. Arslanoğluc, Z. Seda Elmacıc, F. Korela, A. Emrah Çetind, Incorporation of partially purified hen egg white lysozyme into zein films for antimicrobial food packaging, Food Res. Int. 39 (1) (2006) 12–21.
[86] S. Gemili, A. Yemenicioğlu, S.A. Altinkaya, Development of cellulose acetate based antimicrobial food packaging materials for controlled release of lysozyme, J. Food Eng. 90 (4) (2009) 453–462.

[87] C.G. Otoni, S.F. Pontes, E.A. Medeiros, Nde F. Soares, Edible films from methylcellulose and nanoemulsions of clove bud (Syzygium aromaticum) and oregano (Origanum vulgare) essential oils as shelf life extenders for sliced bread, J. Agric. Food Chem. 62 (22) (2014) 5214–5219.
[88] M.H. Hosseini, S.H. Razavi, M.A. Mousavi, Antimicrobial, physical and mechanical properties of chitosan-based films incorporated with thyme, clove and cinnamon essential oils, J. Food Process. Preserv. 33 (6) (2009) 727–743.
[89] H.P. Le, Progress and trends in ink-jet printing technology, J. Imaging Sci. Technol. 42 (1) (1998) 49–62.
[90] K. Suganuma (Ed.), Introduction to Printed Electronics, vol. 74, first ed., Springer, New York, 2014 (Chapter 2).
[91] J. Dixon, Packaging materials: 9 multilayer packaging for food and beverages, Report commissioned by the ILSI Europe Packaging materials task force, Brussels, 2011. Available from: http://ilsi.eu/wp-content/uploads/sites/3/2016/06/ILSI-11-011-9-pack-03.pdf
[92] J. Pan, G. L. Tonkay, A. Quintero, Screen printing process design of experiments for fine line printing of thick film ceramic substrates, Proceedings of International Symposium on Microelectronics, San Diego, CA, USA, November 1–4, 1998, pp. 264–269.
[93] R. Olsson, Some aspects on flexographic ink-paper and paperboard coating interaction, Dissertation, Faculty of Technology and Science Chemical Engineering, Karlstad University, Sweden (2007) 1–66.
[94] J. Sutter, V. Dudler, R. Meuwly, Packaging Materials 8, Printing Inks for Food Packaging Composition and Properties of Printing Ink, Report commissioned by the ILSI Europe Packaging materials task force, Brussels, 2011. Available from: http://ilsi.eu/wp-content/uploads/sites/3/2016/06/ILSI-11-015-printing-inks_finallr.pdf
[95] M. Singh, H. M. Haverinen, P. Dhagat, Ink-jet printing—process and its applications, Adv. Mater. 2010, 22, 673–685.
[96] American Label Group Ltd, Explore the flexographic printing process. Available from: http://www.americanlabel.com/algweb/alg2/explore_flexo.asp
[97] Teuvo Leppanen, Rotogravure printing, UPM RC (C) 2011.
[98] Discovery flexibles, Rotogravure process. Available from: http://www.discoveryflexibles.com/rotogravure/
[99] Offset printing technology, offset lithography, 2016. Available from: http://www.offsetprintingtechnology.com/sub-categories/offset-lithography/
[100] R. De Mondt, Marc Graindourze, UV-Ink-jet printing on food packaging: state of the art and outlook, UV+EB Technology, Radtech International, 2016. Available from: http://www.uvebtech.com/stories/091115/uv-inkjet-printing-food-packaging.shtml#.WB44cup77IU
[101] M. Zalkovskij, L.H. Thamdrup, K. Smistrup, T. Andén, A.C. Johansson, N.J. Mikkelsen, M. Ha. Madsen, J. Garnæs, T.T. Kristiansen, M. Diemer, M. Døssing, D. Minzari, P.T. Tang, A. Kristensen, R. Taboryski, S. Essendrop, T. Nielsen, B. Bilenberg, Smart plastic functionalization by nanoimprint and injection molding, in: D.J. Resnick, C. Bencher (Ed.), Proceedings from SPIE 9423, Alternative Lithographic Technologies VII, vol. 94230T, San Jose, CA, USA, 2015.
[102] Qenos Pty Ltd, Extrusion Coating & Lamination, Quenos technical guide, Polyethylene Technical Guide Series, Victoria, Australia, 2015, pp. 1–24.
[103] The European Commision, Commission recommendation of 18 October 2011 on the definition of nanomaterial, Off. J. Eur. Union (2011) 2011/696/EU, L275/38-L275/40.
[104] J.C. Hannon, J. Kerry, M. Cruz-Romero, M. Morris, E. Cummins, Advances and challenges for the use of engineered nanoparticles in food contact materials, Trends Food Sci. Technol. 43 (2015) 43–62.
[105] S.D.F. Mihindukulasuriya, L.-T. Lim, Nanotechnology development in food packaging: a review, Trends Food Sci. Technol. 40 (2014) 149–167.
[106] D.A. Pereira de Abreu, J.M. Cruz, P. Paseiro Losada, Active and intelligent packaging for the food industry, Food Rev. Int. 28 (2) (2012) 146–187.
[107] X.E. Li, A.N.M. Green, S.A. Haque, A. Mills, J.R. Durrant, Light-driven oxygen scavenging by titania/polymer nanocomposite films, J. Photochem. Photobiol., A 162 (2004) 253–259.
[108] C. Maneerat, Y. Hayata, Gas-phase photocatalytic oxidation of ethylene with TiO_2-coated packaging film for horticultural products, Trans. ASABE 51 (2008) 163–168.

[109] A. López-Gómez, Cerdán-Cartagena, J. Suardíaz-Muro, M. Boluda-Aguilar, M.E. Hernández-Hernández, M.A. López-Serrano, J. López-Coronado, Radiofrequency identification and surface acoustic wave technologies for developing the food intelligent packaging concept, Food Eng. Rev. 7 (2011) 11–32.
[110] A. Mills, D. Hazafy, A solvent-based intelligence ink for oxygen, Analyst 133 (2008) 213–218.
[111] L. Yang, Y. Li, Simultaneous detection of *Escherichia coli* O157:H7 and *Salmonella Typhimurium* using quantum dots as fluorescence labels, Analyst 131 (2006) 394–401.
[112] A. Gnach, A. Bednarkiewicz, Lanthanide-doped upconverting nanoparticles: merits and challenges, Nano Today 7 (2012) 532–563.

CHAPTER NINE

Ensuring Food Safety: General Principles for Safeguarding What You Eat Including the Role of Food Labels

Diana M. Bowman*, Karinne Ludlow**
*Arizona State University, Phoenix, AZ, United States
**Monash University, Clayton, VIC, Australia

Contents

9.1 INTRODUCTION

For many of us, food is much more than simply a source of fuel. It is, instead, multivalued being inextricably linked to custom and culture [1], rituals and religion [2], and family and friends [3]. Whether the food that you buy is, for example, organic or farm to table, contains genetically modified (GM) organisms, or comes from the grocery store, there is an implicit assumption that the food is safe for consumption; that checks and balances, set down by the State in some form, exist so as to ensure safety. This is a reasonable assumption.

However, high profile food scares and controversies—bovine spongiform encephalopathy (BSE) in beef, outbreaks of salmonella in peanut butter, and the United Kingdom's "horse-meat-gate" to name just a few [4–6]—illustrate the very real daily challenges faced by food regulatory schemes. There is a very real human and economic cost to such food safety incidents. Hussain and Dawson [5:586] estimate, for example, that food-borne illnesses are responsible for approximately "300,000 hospitalizations and 5000 deaths in the [United States] alone" annually. The direct and indirect costs of such incidents have been estimated to be $US7 billion per year to the US economy alone [5]. As such, ensuring food safety is big business.

Food safety regulators also face the challenge of overseeing and ensuring safety of new technologies and technological processes within the agri-food sector. These have included the development of GM foods, remote monitoring and sensing technologies, food irradiation, and active packaging materials [7–9]. Such innovation is driven by a range of different demands including: undernourishment and malnutrition, food security concerns, changing weather patterns, increasing globalization of food markets, and economic pressures, to name just a few [10–12]. These emerging technologies can, however, give rise to the so-called "pacing problem," in which the regulator does not possess the necessary legal oversight to adequately regulate the technology, or its products, as they enter the market [13]. These pacing problems can be further exacerbated when the oversight architecture involves multiple tiers, a myriad of actors and numerous regulatory instruments as is the case with the agri-food sector.

The purpose of this chapter is to explore the design and function of the food safety regulatory regime, and the ways in which the regulatory community is responding to one emerging technology which falls within their remit: nanotechnologies. With a growing body of literature now dealing with the actions and approaches being taken by national regulatory agencies, such as the United States (US) Food and Drug Administration (FDA), Food Standards Australia New Zealand and the European Food Safety Authority (EFSA) (see, e.g., [14–16]), the focus of this chapter is on the global food safety architecture, and its application to foods containing nanomaterials. In this vein we look at the global standards that apply to the foods themselves, as well as standards that have been created and adopted for the labeling of such foods. This chapter will focus on nanofoods, including their labeling but will not consider contact materials (see, instead, [17,18]).

Regulation around food safety is called on to address hazards arising from many different sources including chemical, biological, fraud, food-borne, and packaging related [19]. Expectations of regulation of nanobased products are no different in this matter. However, importantly the introduction of nanobased products is often driven as a response to address these hazards in other foods. For example, nanosensors are used to detect pathogens or contaminants in food and nanomaterials are used to improve water filtration, improving food safety.

As other chapters in this book highlight, nanomaterials and nanotechnology-based food processes are increasingly being used within the food sector as a way, for example, within the product itself to enhance aesthetic appeal or within food contact materials to extend a product's shelf life (see also [20,21]). Such products are just the beginning with industry experts suggesting that the next generation of nanobased foods will be on the supermarket shelves within the next few years [22,23].

The development and entry into the market of nanobased foods has not been without controversy; a number of commentators have voiced their concern about the potential risks posed by the use of nanomaterials in food stuffs and within the matrix of food contact materials (see, e.g., [24–26]. It has been argued, for example, that the current regimes and the tools that underpin them are not adequate for safeguarding human and environmental safety, and that nanospecific regulations and tools are needed to ensure safety [26–28]—a classic example of the pacing problem. Commentators have also called for the implementation of a labeling scheme for foods containing nanomaterials so that consumers may exercise an "informed choice" in their purchasing [26,29,30].

Such views are not, however, universally held. Many policy makers, regulators, members of the scientific community, and other key stakeholders have argued that the current regulatory approach is adequate for ensuring consumer safety at this time (for an overview of these discussions, [31,32]). Many have also argued against the introduction of a nanolabel. The complexities of any such label are captured eloquently by Throne-Holst and Rip [33]. Here they reference concerns regarding low consumer awareness of what "nano" is and the subsequently meaningless nature of any such label, as well as the traditional nature of labels to "warn" consumers of a potential hazard. Here the label would be simply to inform of the presence of the nanomaterial, and not of potential risks, which could be confusing to consumers [33].

Although it appears unlikely that these opposing views on safety and labeling will be reconciled anytime soon, an increasing number of nanobased products are making their way into the market, and onto the shelves. Whether they know it or not, consumers will increasingly be purchasing and consuming products containing nanomaterials or foods produced using nanotechnology.

To examine how regulatory regimes have responded to nanobased food products to ensure food safety it is important to first understand the framework in which they operate and the general principles governing their operation. The second part of this chapter provides an overview of this global architecture and the key principles underpinning it. These standards and principles are then incorporated and reflected in the national schemes through their vested powers and regulatory tools. With the question of labeling continuing to be a contentious issue among stakeholders the third part of this chapter provides an in-depth examination of the so-called nanolabeling debate, framed by the general principles outlined in the second part. The chapter concludes by reminding us of the real value of food, and the myriad of challenges it presents to policy makers and regulators alike. Regardless of the technology used to produce it.

9.2 ARCHITECTURE OF INTERNATIONAL FOOD REGULATORY SYSTEMS & GENERAL PRINCIPLES UNDERPINNING THEM

Although there are international agreements addressing food safety, as yet no such agreement expressly addresses nanofoods. Nevertheless, the key general principles created by international regulatory systems will apply to nanofoods and these are explained subsequently. Further, in the context of safety, work is being done by international organizations, such as the Organisation for Economic Co-operation and Development (OECD), which is developing a Consensus and Risk Assessment Document for manufactured nanoparticles and nanofoods.

9.2.1 Codex Alimentarius Commission

Established in 1963 under the auspices of the World Health Organisation (WHO) and the Food and Agriculture Organization (FAO) of the United Nations, Codex Alimentarius Commission (Codex) is the intergovernmental body charged with developing international food standards, principles, and guidelines for the purposes of "protecting consumers' health and ensuring fair practice in food trade" [34:5, Part I (1)]. As of December 2016, 187 countries were considered to be Member Countries of Codex, with the EU being the only Member Organization. As noted on the Codex website, these Members are home to "99% of the worlds population" [35]. The body has a further 234 Observer Members, which include international governmental organizations, nongovernmental organizations, and United Nations Organizations [35]. Together, these Members shape the agenda of Codex, and provide the necessary expertise and assistance to enable the development of scientifically based standards, guidance, and other documents that become the reference points for food safety across the globe.

General and specific standards, along with guidelines, codes of practice and recommendations produced by Codex are published in the Codex Alimentarius, or "Food Code." As of November 2015, the Alimentarius consisted of 341 documents, including 212 standards, 73 guideline documents, and 51 codes of practice [36]. These include, for example, the *General Standard for Food Additives* (*Codex Stan 192-1995*), the *Standard for Chocolate and Chocolate Products* (*Codex Stan 87-1981, Rev. 1-2003*), and the *Guideline for the Production, Processing, Labelling and Marketing of Organically Produced Foods* (*GL 32-1999*). These documents are produced through a rigorous, consensus driven, process.[1] As such, it is not uncommon for the process to take years from start to finish.

Importantly, the standards, guidelines, and other documents produced by Codex are not binding on Member Countries. Member Countries are, instead, strongly encouraged to incorporate the standards and other materials published in the Alimentarius into their domestic regulatory frameworks as a way to promote food safety, increase harmonization, and promote trade [37]. Where a standard is a "minimum standard," a Member Country retains the flexibility to adopt more stringent standards to protect consumer safety (when based on a scientific basis) [38]. This provides Member Countries with a degree of flexibility in defining the level of protection that they wish to implement within their own jurisdiction. However, as noted subsequently, even this is subject to further restraints.

Given the scope of Codex's work, there can be little doubt that food incorporating nanomaterials or nanoscale food additives falls within their remit. Currently though, there is no express reference to such foods or additives in the Alimentarius. Should the Member Countries wish to expressly bring these issues within the body's work a Member Country would need to put forward a proposal to the Executive Committee, in the manner set out in the Procedural Manual [39]. This by itself would not guarantee that the proposal would go forward, with the Executive Committee having to assess each proposal against a number of specific criteria. This includes, for example, relevance to their strategic plan and the needs of developing countries [39].

One area in which the Executive Committee may receive a request for new work would be in the area of developing a labeling standard for foods containing nanomaterials. As shall be discussed in more detail in the conclusion, the EU has already moved ahead with requiring foods containing "engineered nanomaterials" (as defined in *Regulation (EU) No 1169/2011)* to list nanoscale ingredient in a food's list of ingredients (Article 18(3)). Such a work item would not be without precedent with Codex having previously commenced deliberations for a labeling standard for GM food [40].

The GM labeling deliberations provide a case-in-point as to the challenges that a Member Country may face should it push forward with a similar work item for the labeling of nanobased foods. Deliberations over GM labeling lasted well over a decade due to highly polarized views on the need for, and nature of, any such label [41,42]. Key issues of debate and disagreement included, for example, the practical implications of a labeling standard, the "product v process" dichotomy underpinning different regulatory approaches, a lack of scientific risk analysis and toxicological studies, economic costs, and consumer rights [37,40]. "With the same arguments presented *ad nauseam*" and a "lack of consensus about compulsory labeling for foods derived from modern biotech" [43:971], a decision was reached to discontinue the work program in relation to definitions and detailed rules. In its place, Member Countries agreed to the adoption of the *Compilation of Codex Texts Relevant to Labeling of Foods Derived from Modern Biotechnology* (*CAC/GL 76-2011*). The purpose of the document, published by the Codex Committee on Food Labelling is to simply,

> *"recall and assemble in a single document some important elements of guidance from Codex texts, which are relevant to labelling of foods derived from modern biotechnology" (Article 1).*

This amounts to ten different standards ranging from, for example, the *General Standard for the Labelling of Prepackaged Foods* (*Codex Stan 1-1985*) through to the *Guidelines for Use of Nutrition and Health Claims* (*CAC/GL 23-1997*), and the *Working Principles for Risk Analysis for Food Safety for Application by Governments* (*CAC/GL 62-2007*).

Importantly, for the purposes of this chapter, *CAC/GL 76-2011* "is not intended to suggest or imply that foods derived from modern biotechnology are necessarily different from other foods simply due to their method of production" [44]. Codex also adopted principles for risk analysis for food safety for foods derived from genetic modification, which establish that if a risk is identified, labeling is an appropriate management strategy [Principles for the Risk Analysis of Foods Derived from Modern Biotechnology (*CAC/GL 44-2003*:1)].

In our view, should Codex pursue a labeling standard for foods containing nanomaterials, debates analogous to those that played out over a decade or two in relation to GM foods are likely to occur. A similar stalemate and/or lack of consensus on any such standard would be the most likely outcome.

In regard to food safety, the general principles of Codex and national regulatory systems will apply to nanofoods as they do to other foods. This includes the principle that new food substances shall be subject to independent and scientifically based safety assessment prior to sale. If a food is not assessed as safe, it will not be approved for sale at the national level. As a general rule, commercial sponsors of the new food substance will have the burden of demonstrating safety in accordance with specified testing requirements. Such sponsors will also need to ensure that there are well-defined intended uses (including the level of use and foods in which it will be used) and compositional specifications of added substances. These definitions are central to the development of scientific data (such as acceptable intake levels) on the safety of the food and verification of that data as well as to minimize fraudulent and misleading uses.

9.2.2 World Trade Organization

As noted earlier, member countries are not bound to adopt the standards produced by Codex. Although this is true, the reality is much more complex given the relationship between Codex and the World Trade Organization (WTO), and the various binding instruments that fall under the latter's umbrella. Two key instruments administered by the WTO—the *Sanitary and Phytosanitary Measures (SPS) Agreement* and the *Technical Barriers to Trade (TBT) Agreement*—specifically refer to, and rely upon, the international food standards, guidelines, and recommendations produced by Codex as the benchmark for food safety.

By way of background the WTO was established on 1 January 1995 as a replacement body to the Contracting Parties of the General Agreement on Tariffs and Trade (GATT). The WTO is the key international institution concerned with the liberalization of trade rules (as provided for under the World Trade Organization Agreement (WTOA)) [45]. As such, the WTO has an interest in the governance of all nanotechnology products, including nanofoods, in so far as domestic regulatory frameworks impact on the trade of the products. WTO's primary concern shall, however, be in relation to whether or not the regulatory approach adopted by a Member Country for a nanobased product results in the restriction of trade.

As with GM foods, any safety concerns raised by WTO members in relation to nanofoods would fall within the scope of the *SPS Agreement*. As stated in Article 1—General Provisions, the *SPS Agreement* "applies to all sanitary and phytosanitary measures which may, directly or indirectly affect international trade." Under the SPS Agreement, any SPS measure implemented by a WTO Member Country[2] which conforms with a Codex standard will be *prima facie* consistent with that nation's obligations under the Agreement.

However, in the absence of any Codex standards, guidelines, and/or recommendations that deal specifically with nanofoods governments will be required to base any national measures on either analogous international standards [38], or demonstrate that the higher level protection is based on scientific risk assessments to the extent necessary to protect human and environmental safety (*SPS Agreement*, Article 5.1[3]). The establishment of a higher level of protection will need to be deemed to be scientifically justifiable, if it is to survive its examination by other member states [37].

Given the lack of toxicological and epidemiological studies reported on potential human health risks associated with the ingestion of foods containing nanomaterials, in conjunction with contradictory findings, it appears likely that members may resort to relying on Article 5.7[4] of the *SPS Agreement* when formulating national SPS measures. However, although Article 5.7 enables a member state to employ provisional SPS measures under strict conditions, as established by the Appellate Body in *Japan–Measures Affecting Agriculture Products (Japan–Agricultural Products)*[5], the decision of the Appellate Body in *EC-Biotech Products*[6] suggests that it will be extremely difficult for a member to have a right to recourse under Article 5.7 when formulating their national SPS frameworks. Further, provisional measures introduced by a Member under Article 5.7 must be reviewed within a reasonable period by seeking additional information to conduct an objective risk assessment [46:27].

Moreover, should a Member wish to incorporate a precautionary approach to nanofoods within their national SPS framework, such measures would need to be consistent with existing scientific evidence in order for the member to act consistently with its obligations under the *SPS Agreement*. This is because, as noted by Gonzalez [47], Article 5.7 is triggered by insufficiency of scientific evidence and not by scientific uncertainty.

In contrast, any "technical regulations and standards, including the packaging, marking, and labelling requirements" (Preamble, *TBT Agreement*) concerns raised by WTO members in relation to nanofoods that are not covered by the SPS Agreement will fall within the scope of the TBT Agreement. In doing so, the Agreement further recognizes, however, that "no country should be prevented from taking measures necessary to ensure the quality of its exports, or for the protection of human, animal, or plant life or health..." (Preamble, *TBT Agreement*). Although the TBT Agreement does not identify relevant standard setting bodies for international standards, as with the SPS Agreement the establishment by a member state of a higher level of protection in relation to nanofoods would need to be based on scientific risk assessment data. As such, if a country implemented, for example, a national labeling policy for nanofoods for reasons other than food safety, such an action could, potentially, be challenged by another country on the basis that the policy was an unnecessary barrier to trade.

9.2.3 Impact at the national and regional level

National and regional regulatory systems for nanofoods will be subject to the constraints imposed by international agreements, such as those described earlier. Within that constraint, although the food safety principles and guidelines for implementation provide the foundation for oversight at regional and national levels, countries have responded differently in the manner of codification and application to nanofoods.

In most cases, the broad policy principles and general approach—such as the requirement for premarket assessment and review to assure safety—are mandated in laws enacted by parliaments or other legislative bodies. Critical scientific principles and rules for conducting safety assessments are typically codified in regulations or less formal directives or guidance documents issued by bodies, such as the European Food Safety Authority (EFSA)[7] and the US FDA.[8]

Other relevant internationally applicable general principles include the Codex *Code of Ethics for International Trade in Food (CAC/RCP 20-1979)* and *Principles for Food Import and Export Inspection and Certification (CAC/GL 20-1995)*. Article 4.2 of *CAC/RCP 20-1979* provides that no food should be in international trade which, inter alia, has in it any substance in an amount which renders it harmful or injurious to health, consists of foreign matter or is otherwise unfit for human consumption or is adulterated. This would apply to nanofoods if they meet any of those characteristics.

As noted earlier, a central regulatory principle governing all new food substances, whether nanoform or not, is that they undergo independent, scientifically sound safety assessment prior to marketing. Pursuant to section 1.1 of the *General Standard for Food Additives (Codex Stan 192-1995)* only listed food additives are recognized as suitable for use in foods and only additives that meet premarketing safety assessment can be listed. This approach is adopted by national regulatory systems, such as the *Australia New Zealand Food Standards Code–Standard 1.3.1–Food Additives*, clause 3. One difficulty for national regulatory systems is where a nanoform of a previously approved substance is introduced. If such a material is not considered "new" it will not be subject to reevaluation.

Other key general principles regarding safety include that each nation can establish its own "appropriate level of protection" (ALOP) for food substances. Most national and regional regulatory systems do this by codifying a protective safety standard for intentionally added substances. For example, the US FDA uses "safe" as its standard. Safe (or safety) is defined in *21 Code of Federal Regulations* (CFR) § 170.3(i)) as to mean:

> *"that there is a reasonable certainty in the minds of competent scientists that the substance is not harmful under the intended conditions of use. It is impossible in the present state of scientific knowledge to establish with complete certainty the absolute harmlessness of the use of any substance. Safety may be determined by scientific procedures or by general recognition of safety...."*

The burden of proof on safety falls on the commercial sponsor. Nevertheless, the commercial sponsor must, when conducting the safety testing, proceed in accordance with specified testing requirements and demonstrate safety. If safety cannot be demonstrated, approval in the form of listing will be denied. Further *CAC/GL 20-1995* requires countries to keep up to date with available scientific evidence. For example, section 3.6 requires inspection and certification procedures to be fully effective in achieving their designated objectives having regard to the determination of acceptable levels of protection which is required and section 3.7 requires risk assessment to be based on current available scientific evidence (for a discussion on the current scientific evidence, and questions regarding risks for nanomaterials in foods, please refer to chapters [Chapter 5] and [Chapter 6]).

9.3 LABELING

It is generally recognized that food labeling, as described by Codex, "is the primary means of communication between the producer and seller of food on the one hand, and the purchaser and consumer on the other" [48:i]. As such, food labeling may provide a benefit to both the consumer and the seller as it can assist the consumer in making an informed decision on the product, while promoting the movement of goods within the market [49].

The internationally accepted general principles relating to the labeling of prepackaged foods are set down in Codex *General Standard for the Labelling of Prepackaged Foods* [*Codex Standard 1-1985 (Rev. 1-1001)*]. Pursuant to section 2 of the *General Standard*, "labeling" includes,

> *"any written, printed or graphic matter that is present on the label, accompanies the food, or is displayed near the food, including that for the purpose of promoting its sale or disposal".*

A "label" is defined in section 2 to mean "any tag, brand, mark, pictorial or other descriptive matter, written, printed, stenciled, marked, embossed or impressed on, or attached to, a container of food." As such, as eloquently noted by Einsiedel [50:231], the definition of a label as supplied by Codex is broad enough to include simple symbols or logos, as well as "something as complex as a set of ingredients (with a string of chemical information) and nutritional information."

In this part of the chapter we focus on the internationally accepted principles relating to the regulation of food labeling and their application to foods containing nanomaterials. Where appropriate, we then consider their application in specific jurisdictions to illustrate how they may be applied within a national context.

9.3.1 Considerations in relation to the labeling of prepackaged foods

As noted earlier, the internationally accepted principles relating to prepackaged foods labeling are set down in *Codex Standard 1-1985*. These are reflected in legislative instruments of participating jurisdictions including, for example, the EU under *Regulation (EU) No. 1169/2011 on the provision of food information to consumers* and the US under *Title 21 (Food and drugs), Chapter I, Subchapter B (§101) of the Federal Food, Drug and Cosmetic Act*.

Pursuant to section 1 of *Codex Standard 1-1985* the,

> *"standard applies to the labelling of all prepackaged foods to be offered as such to the consumer or for catering purposes and to certain aspects relating to the presentation thereof".*[9]

A general principle embodied in section 2 of the *Standard* is that labeling must not be "false, misleading, or deceptive." Mandatory information to appear on the label of prepackaged foods is set out in section 4 and includes, but is not limited to, the name of the food, list of ingredients which are to appear in descending order of ingoing weight at the time of manufacturing and food additives which serve a technological function. Pursuant to section 5.1, the label must also provide a quantitative list of ingredients by weight or volume. Additional information may be voluntarily incorporated onto the label as long as it does not conflict with the requirements set out in *Codex Standard 1-1985* (section 7). *Prima facie*, these principles apply equally to prepackaged foods that do, and do not, contain nanomaterials.

As provided for under section 4.1, the name of the food must be indicated on the label, in a form that indicates the "true nature of the food and normally be specific and not generic." Pursuant to section 4.1.2,

> *"there shall appear on the label either in conjunction with, or in close proximity to, the name of the food, such additional words or phrases as necessary to avoid misleading or confusing the consumer in regards to the true nature and physical condition of the food including but not limited to the type of packaging medium, style, and the condition or type of treatment it has undergone...".*

For example, fruit that has been sliced must be labeled as such to inform the consumer as to the exact nature of the prepackaged food. As the label must not mislead the consumer as to the "true nature" of the food, a fundamental question here in relation to foods processed with or containing nanomaterials is whether or not the condition or type of treatment it has undergone is sufficient to render the food different from its conventional food counterpart. If the true nature and physical condition of a food containing nanomaterial differs enough from that of a conventional food, the argument could be made under this section that the label on the food must contain "additional words or phrases as necessary to avoid misleading or confusing the consumer." Any such conclusion would have to be made on a product-specific basis.

Section 4.2 of *Codex Standard 1-1985* requires a listing of ingredients. Such ingredients are required to be listed on the basis of their common or usual name (specific or class name) as a means to inform the consumer of the type and form of the prepackaged food; the principle does not require information relating to the size of the ingredient present in the food to be included as part of the listing requirement.[10]

It is arguably not surprising, therefore, that with the exception of the EU, jurisdictions have not sought to differentiate conventional ingredients from those at the nanoscale for the purposes of complying with section 4.2 of the *Standard*. The EU has, however, done exactly that. Pursuant to Article 18(3) of *Regulation (EU) No. 1169/2011*,

> *"All ingredients present in the form of engineered nanomaterials shall be clearly indicated in the list of ingredients. The names of such ingredients shall be followed by the word 'nano' in brackets".*[11]

The purpose of this provision, as stated in the Preamble, is to "inform consumers of the presence of engineered nanomaterials in food" (Preamble, 25). Let's take an example. As noted by Weir et al. [51], titanium dioxide (TiO_2) has been a common ingredient in food products for the purposes of "whitening" the food. The white powder, commonly referred to as E171 in the EU[12] or food grade titanium dioxide can be found in products as diverse as mozzarella cheese, donuts, and chewing gum [51,52]. As reported by Weir et al. [51], approximately 35% of the TiO_2 particles in the food stuff shall be at the nanoscale. As such, the ingredients list for a chewing gum, for example, sold in the EU would now need to indicate this. The ingredient list would therefore look something like this: sorbitol, gum base, natural and artificial flavorings, glycerol. Less than 2% of: soy lecithin, aspartame-acesulfame, aspartame mannitol, colors (titanium dioxide (nano), red 40).

In this vein, and articulated in the Preamble of *Regulation (EU) No. 1169/2011* the labeling obligation provides consumers with an environment of greater transparency; it is not their intent for the "nano" listing to be read as a warning. Although this requirement may have been driven by a wish to "inform consumers about nanomaterials in food" (Preamble at 25) it also provides the regulator with an additional tool for monitoring, and tracing, the use of nanomaterials within the EU market.

Section 4.2.4 requires processing aids and food additives to be included in the list of ingredients under certain conditions; listing is done on the basis of the name of the processing aid or food additive. The requirement for listing of such a food additive is based on it being present in a "significant quantity or in an amount sufficient to perform a technological function." As such, nanoscale food additives which are carried over into foods may be required to be listed on the label, even if present in small quantities but in an amount sufficient to perform a technological function. However, even if they are listed, the label does not need to provide information on the size of the food additives; rather only the name of the food additive must be listed.

Importantly, it would appear that *Codex Standard 1-1985* does not prevent the voluntary labeling of prepackaged foods that contain nanomaterials (either positive or negative), as long as such information is not in conflict with the *Standard* generally, including being "false, misleading, or deceptive." As such, it can be concluded that *Regulation (EU) No. 1169/2011* is in compliance with the *Standard*.

9.3.2 Considerations in relation to labeling of additives

In addition to the requirements relating to the labeling of food additives set out in section 4.2.4 of *Codex Standard 1-1985*, internationally recognized standards in relation to the labeling of food additives[13] are set down in the *General Standard for the Labelling of Food Additives When Sold As Such (Codex Standard 107-1981).*

As with *Codex Standard 1-1985*, a general principle embodied in *Codex Standard 107-1981* is that labeling must not be "false, misleading, or deceptive" (section 3). Pursuant to section 4.1(a) of *Codex Standard 1-1985*,

> *"the name of each food additive present shall be given. The name shall be specific and not generic and shall indicate the true nature of the food additive…"*

This requirement would appear to apply equally to food additives regardless of the particle size of the food additive, as it refers to a listing on the basis of name and not size. Within the EU, pursuant to *Regulation (EU) No. 1169/2011*, any food additive at present within the food at the nanoscale (as defined by the Regulation) would need to be labeled as such.

Arguably, however, the term "true nature" as it is stated here is ambiguous. Outside of the EU, a stakeholder could argue that the term should be read broadly so as to encompass the particle size of the food additive. Thus, nanoscale food additives would have to be labeled as such. Whether or not such an argument is successful depends on the interpretation of the term "*true nature*" within the overall context of the *Codex Standard 107-1981,* which does not refer to physico-chemical characteristics elsewhere in its text.

9.3.3 Considerations concerning claims

The commonly defined internationally accepted principles relating to claims for foods may be found in several Codex Guideline and Standard documents. The most relevant to this chapter include the *General Guidelines on Claims (CAC/GL 1-1979*) (*Rev. 1-1991)* and *Guidelines for Use of Nutrition and Health Claims (CAC/GL 23-1997).* These two documents, and how they relate to the labeling of foods containing nanomaterials, are discussed in turn.

The *General Guidelines on Claims (CAC/GL 1-1979)* sets out the accepted principles relating to "claims made for food irrespective of whether or not the food is covered by an individual Codex Standard" (section 1.1). A general principle embodied in *CAC/GL 1-1979* is that "no food should be described or presented in a manner that is false, misleading or deceptive or is likely to create an erroneous impression regarding its character in any respect" (section 1.2). For the purposes of the *Guideline*, a claim is defined as including "any representation which states, suggests, or implies that a food has particular characteristics relating to its origin, nutritional properties, nature, production, processing, composition, or any other quality" (section 2). Any claim made by an individual must be able to be substantiated in order for it to be permissible (section 3.3). The principles set out in *CAC/GL 1-1979* will apply equally to claims made in relation to foods that do and foods that do not contain nanomaterials.

As an example, if a person marketing a food wishes to make the claim that the food is processed using nanotechnologies (a production related claim) then they must be able to substantiate this claim in order for it to be an admissible claim. Any such claim must also conform with all other requirements set out in *CAC/GL 1-1979* as well as any other relevant guidelines.

The *Guidelines for Use of Nutrition and Health Claims (CAC/GL 23-1997)* set down the commonly accepted principles for nutrition and health claims in food labeling. As stated in its Preamble, the nutrition and health claims should be consistent with, and support, national nutrition and health policies. Health claims, in particular,

> *"should be supported by a sound and sufficient body of scientific evidence to substantiate the claim, provide truthful and non-misleading information to aid consumers in choosing healthful diets…" (Preamble).*

The scope of *CAC/GL 23-1997* is set out in section 1.1 so as to "relate to the use of nutrition and health claims in food labelling and, where required by the authorities having jurisdiction, in advertising".[14] Pursuant to section 3,

> *"any food for which a nutrition or health claim is made should be labelled with a nutrient declaration in accordance with section 3 of the Codex* Guidelines on Nutritional Labelling*"*.

Only such nutrition claims which relate to,

> *"energy, protein, carbohydrates, and fats and components thereof, fibre, sodium and vitamins and minerals for which Nutrient Reference Values have been laid down…",*

are permissible under section 4.1 of *CAC/GL 23-1997*. Permitted nutrient claims on the label would include, for example, the phrase "a low fat food." These principles concerning the labeling of nutrition claims, including comparative claims, apply to all foods to which nutrition claims are made and will apply equally to foods that do and do not contain nanomaterials. At this time it appears unlikely that a manufacturer of foods containing nanomaterials would or, indeed, could make any such claims.

The requirements for health claims are set out in section 8 of the *CAC/GL 23-1997*. As provided by section 8.1, health claims are permitted as long as they are,

> *"based on current relevant scientific substantiation and the level of proof must be sufficient to substantiate the type of claimed effect and the relation to health as recognized by generally accepted scientific review of the data and the scientific substantiation should be reviewed as new knowledge becomes available…" (section 8.1.1).*

Moreover, pursuant to section 8.1.2, "any health claim must be accepted by or be acceptable to the competent authorities of the country where the product is sold." An example of a health claim, as provided in section 2.2.1 of the *Guidelines,* is as follows:

> *"Nutrient A (naming a physiological role of nutrient A in the body in the maintenance of health and promotion of normal growth and development). Food X is a source of/high in nutrient A".*

Any such claim must be truthful and supported by relevant scientific evidence. These principles concerning the labeling of health claims, as well as those relating to healthy diets, apply to all claims made in relation to foods, and will apply equally to foods that do and do not contain nanotechnologies. At this time, at least to our knowledge, no national regulator has accepted a health claim specifically associated nanomaterials. As such, it appears unlikely that such a claim could, or would, be made by a manufacturer in the near future.

9.3.4 Considerations in relation to nutrition labeling

The internationally accepted principles relating to nutrition labeling for all foods are set down in the *Guidelines on Nutritional Labelling (CAC/GL 2-1985) (Rev. 1–1993)*. The purpose of *CAC/GL 2-1985* is "to ensure that no nutritional claims are made without nutritional labeling" (Preamble). As with other Standards and Guidelines examined in this section of the chapter, a general principle embodied in *CAC/GL 2-1985* is that the nutritional labeling is not in anyway "false, misleading, or deceptive."

Pursuant to section 2.1 of *CAC/GL 2-1985*, nutrition labeling "is a description intended to inform the consumer of nutritional properties of food." A nutritional claim,

> *"means any representation which states, suggests or implies that a food has particular nutritional properties including but not limited to the energy value and to the content of protein, fat and carbohydrates, as well as the contents of vitamins and minerals..." (section 2.4).*

Where a nutritional claim is made, section 3 of the *Guidelines* mandates the inclusion of a nutrient declaration on the labeling. Such information must include, but not limited to, energy value, amount of protein, available carbohydrate and fat, and the amount of any other nutrient considered being relevant for a good nutritional status.

These labeling requirements for nutritional labeling will apply to foods that do and do not contain nanomaterials. The *Guideline* does not differentiate between "nutrients" of varying sizes, and as such, it would appear that such labeling does not require the particle size of the nutrient to which the claim refers to be listed. Rather, as required by section 3, listing of nutrients is on the basis of the name of the substance to which the nutritional claim is being made. It does not appear to matter per se if the nutrient may or may not be formulated and present at the nanoscale; it is the amount and type and energy value thereof, for example, that is important and must be declared under the *Guidelines.*

The use of labels as a food safety tool, while a new tool per se, should still fall within the regulations described earlier. For example, where a label monitors for pathogen presence, the label will need to be accurate to avoid the product facing challenges as a misleading claim under *CAC/GL 1-197*.

9.3.5 Other consumer information issues

As noted earlier, the labeling of food products containing nanomaterials is becoming an increasingly controversial issue in a number of jurisdictions. At this stage it would appear that, as with the national labeling policies and approaches that have emerged in relation to genetically modified organisms (GMOs), for example, there is the potential for jurisdictions to adopt one of the three broad approaches:

1. retain the current status quo,
2. pursue voluntary labeling programs, or
3. pursue mandatory labeling programs (the approach adopted by the EU).

Caswell has argued, for example, in relation to GMOs, that "governments are likely to prefer voluntary or mandatory approaches based on their perception of what proportion of their citizens want information about the technology" [53:24]. However, Caswell [54:56] notes that the labeling policy ultimately adopted by governments will be largely driven by "its perception of benefits and costs." Labeling policies for food products containing or processed using nanomaterials will be similarly made on an economic basis, as well as the politics of differing social value systems and consumer preferences. It is these factors that appear to have played a strong hand in the European Parliament's push for nanospecific provisions in *Regulation (EU) No. 1169/2011 on the provision of food information to consumers and Regulation (EU) 2015/2283 on novel foods.* This will have the potential for not only influencing the amount of information consumers have access to in relation to such products, but will—as we have already seen—give rise to divergent labeling policies amongst jurisdictions.

9.4 LOOKING FORWARD

Looking forward returns us to the beginning—food is more than fuel. Nanotechnology can add more value to this already value-laden sector but it, like other innovations, will challenge regulatory regime as it develops. Although not unique to nanofoods, food safety regulation around nanotechnology will be most challenged at, what has been called, the intersections of applications of food beyond its traditional applications, including developments, such as food-mediated targeted nutrient delivery to help maintain or improve health, which will cause food science to merge with nutritional science and the interface of food safety regulation with drug regulation [55:35–36]. The application of food for gastronomy and pleasure through new types of food structures, tastes, and textures may influence the attitudes of some stakeholders to nanofoods. However, in all cases, food safety will be assumed by the public and regulation will be needed to deliver that.

ENDNOTES

[1] The details of which are set out in Codex's Procedural Manual (version 23). Available from: ftp://ftp.fao.org/codex/Publications/ProcManuals/Manual_23e.pdf.

[2] As of December 2016, 164 countries were members of the WTO [45].

[3] Article 5.1 of the SPS Agreement states that, "Members shall ensure that their sanitary or phytosanitary measures are based on assessment, as appropriate to the circumstances, of the risks to human, animal or plant life or health, taking into account risk assessment techniques developed by the relevant international organizations."

[4] Article 5.7 of the SPS Agreement states that, "In cases where relevant scientific evidence is insufficient, a Member may provisionally adopt sanitary or phytosanitary measures on the basis of available pertinent information, including that from the relevant international organizations as well as from sanitary or phytosanitary measures applied by other Members. In such circumstances, Members shall seek to obtain the additional information necessary for a more objective assessment of risk and review the sanitary or phytosanitary measure accordingly within a reasonable period of time."

[5] *Japan—Measures Affecting Agricultural Products*, WT/DS76/AB/R (22 February 1999). In this dispute, the Appellate Body dealt with a complaint by the USA 'relating to the requirement imposed by Japan to test and confirm the efficacy of the quarantine treatment for each variety of certain agricultural products ("the varietal testing requirement")'. In their decision, the Appellate Body stated that Article 5.7 operates as a qualified exemption from the obligation under Article 2.2 not to maintain SPS measures without sufficient scientific evidence. An overly broad and flexible interpretation of that obligation would render Article 5.7 meaningless (at paragraph 80). The Appellate Body subsequently articulated four requirements that must be meet in order for a Member to adopt and maintain a provisional SPS measure: (1) imposed in respect to a situation where relevant scientific information is insufficient; and (2) adopted on the basis of available pertinent information. After adopting provisional measures, the Member must: (3) seek[s] to obtain the additional information necessary for a more objective assessment of risk; and (4) review[s] the ...measure accordingly within a reasonable period of time (at paragraph 89).

[6] *WTO Panel Reports on European Communities-Measures Affecting the Approval and Marketing of Biotech Products* WT/DS291R, WT/DS292R, WT/DS293R 29 September 2006.

[7] See, for example, EFSA, Technical guidance for establishing the safety of additives for the consumer (EFSA-Q-2008-406) (16 September 2008) (Available from: http://www.efsa.europa.eu/EFSA/efsa_locale-1178620753812_1211902094077.htm).

[8] See, for example, FDA/CFSAN/Office of Food Additive Safety, Recommendations for Submission of Chemical and Technological Data for Direct Food Additive Petitions (March 2009) (Available from: http://www.cfsan.fda.gov/~dms/opa2cg5.html).

[9] This general standard is reflected in numerous Codex Standards relating to prepackaged foods including, for example, *Codex Standard for Ginger* (*Codex Standard 218-1999, AMD. 1-2005*) (see section 6.1), *Codex Standard for Asparagus* (*Codex Standard 225-2001, AMD. 1-2005*) (see section 6.1) and the *Codex Standard for Chocolate and Chocolate Products* (*Codex Standard 87-1981, Rev. 1 - 2003*) (see section 5).

[10] This requirement is expressly incorporated into a number of other Codex Standards including, for example, *Codex Standard for Formula Foods for Use in Weight Control Diets (Codex Standard 181-1991*) (see section 9.2).

[11] For the purposes of the Regulation, an 'engineered nanomaterial' is defined in article 2(t) to mean: "any intentionally produced material that has one or more dimensions of the order of 100 nm or less or that is composed of discrete functional parts, either internally or at the surface, many of which have one or more dimensions of the order of 100 nm or less, including structures, agglomerates or aggregates, which may have a size above the order of 100 nm but retain properties that are characteristic of the nanoscale."

[12] As listed in Annex I of *Directive 94/36/EEC.*

[13] For the purposes of Standard 107-1981, a "food additive" is defined in section 2 of the *Standard.*

[14] For the purposes of *CAC/GL 23-1997*, a "nutrition claim" is defined in Section 2.1, and a 'health claim' is defined in section 2.2. "Other function claims" is defined in section 2.2.2.

REFERENCES

[1] P. Fieldhouse, Food and Nutrition: Customs and Culture, Springer, New York, (2013).
[2] S.W. Mintz, C.M. Du Bois, The anthropology of food and eating, Ann. Rev. Anthropol. 31 (2002) 99–119.
[3] A. Murcott, The Sociology of Food and Eating, Gower Publishing Co, New York, (1983).
[4] D. Vogel, What's the Beef?: The Contested Governance of European Food Safety, MIT Press, Boston, (2006).
[5] M.A. Hussain, C.O. Dawson, Economic impact of food safety outbreaks on food businesses, Foods 2 (4) (2013) 585–589.
[6] E.J. Abbots, B. Coles, Horsemeat-gate: the discursive production of a neoliberal food scandal, Food Cult. Soc. 16 (4) (2013) 535–550.
[7] H.W. Kendall, Bioengineering of crops: report of the World Bank Panel on Transgenic Crops, World Bank, Washington, DC, (1997).
[8] N. Wang, N. Zhang, M. Wang, Wireless sensors in agriculture and food industry—recent development and future perspective, Comput. Electron. Agric. 50 (1) (2006) 1–14.
[9] L. Vermeiren, F. Devlieghere, M. Van Beest, N. De Kruijf, J. Debevere, Developments in the active packaging of foods, Trends Food Sci. Technol. 10 (3) (1999) 77–86.
[10] Organization for Economic Cooperation and Development and Food and Agriculture Organization of the United Nations, OECD-FAO Agricultural Outlook 2015, OECD Publishing, Paris, 2015.
[11] J.I. Cohen, Poorer nations turn to publicly developed GM crops, Nat. Biotechnol. 23 (1) (2005) 27–33.
[12] B. Smit, M.W. Skinner, Adaptation options in agriculture to climate change: a typology, Mitig. Adapt. Strateg. Glob. Change 7 (1) (2002) 85–114.
[13] G.E. Marchant, B.R. Allenby, J.R. Herkert (Eds.), The Growing Gap Between Emerging Technologies and Legal-Ethical Oversight: The Pacing Problem, vol. 7, Springer Science & Business Media, New York, 2011.
[14] Food and Drug Administration, Nanotechnology—A report of the US food and drug administration nanotechnology task force, FDA, Washington DC, 2007.
[15] K. Ludlow, D.M. Bowman, G.A. Hodge, Final Report: Review of Possible Impacts of Nanotechnology on Australia's Regulatory Frameworks, Monash Centre for Regulatory Studies, Monash University, 2007.
[16] European Commission, Regulatory Review on Nanomaterials, COM, 366, EC, Brussels, 2008.
[17] Q. Chaudhry, A. Gergely, D.M. Bowman, Regulatory frameworks for food, in: Q. Huang (Ed.), Nanotechnology in the Food, Beverage and Nutraceutical Industries, Woodhead Publishing, Sawston, 2012, pp. 85–101.
[18] D.M. Bowman, Q. Chaudhry, A. Gergely, Evidence-based regulation of food nanotechnologies: a perspective from the European Union and United States, in: C. Sabliov, H. Chen, R. Yada (Eds.), Nanotechnology and Functional Foods: Effective Delivery of Bioactive Ingredients, Wiley-Blackwell, NJ, 2015, pp. 358–374.
[19] F.J. Van de Brug, N.B. Lucas Luijckx, H.J. Cnossen, G.F. Houben, Early signals for emerging food safety risks. From past cases to future identification, Food Control 39 (2014) 75–86.
[20] Q. Chaudhry, M. Scotter, J. Blackburn, B. Ross, A. Boxall, L. Castle, R. Aitken, R. Watkins, Applications and implications of nanotechnologies for the food sector, Food Addit. Contam. 25 (3) (2008) 241–258.
[21] Q. Chaudhry, L. Castle, R. Watkins, Nanotechnologies in Food, Royal Society of Chemistry, London, (2010).
[22] S. Ranjan, N. Dasgupta, A.R. Chakraborty, S.M. Samuel, C. Ramalingam, R. Shanker, A. Kumar, Nanoscience and nanotechnologies in food industries: opportunities and research trends, J. Nanopart. Res. 16 (6) (2014) 1–23.
[23] X. Jiang, D. Valdeperez, M. Nazarenus, Z. Wang, F. Stellacci, W.J. Parak, P. Del Pino, Future perspectives towards the use of nanomaterials for smart food packaging and quality control, Part. Part. Syst. Charact. 32 (4) (2015) 408–416.
[24] G. Scrinis, K. Lyons, The emerging nano-corporate paradigm: nanotechnology and the transformation of nature, food and agri-food systems, Int. J. Sociol. Food Agric. 15 (2) (2007) 22–44.

[25] M. Cushen, J. Kerry, M. Morris, M. Cruz-Romero, E. Cummins, Nanotechnologies in the food industry—recent developments, risks and regulation, Trends Food Sci. Technol. 24 (1) (2012) 30–46.
[26] Friends of the Earth, Tiny Ingredients, Big Risks, FoE, Washington, DC, 2014.
[27] ETC Group, Down on the Farm. The Impacts of Nano-Scale Technologies on Food and Agriculture, ETC Group, Ottawa, 2004.
[28] Friends of the Earth, Nanotechnology Policy Statement, FoE, Melbourne, 2007.
[29] G. Miller, S. Kinnear, Nanotechnology–the new threat to food, FoE, Melbourne, 2008.
[30] ANEC/BEUC, Nanotechnology: small is beautiful but is it safe? Joint ANEC/BEUC position, ANEC/BEUC, Brussels, June 2009.
[31] G.A. Hodge, D.M. Bowman, A.D. Maynard (Eds.), International Handbook on Regulating Nanotechnologies, Edward Elgar, Cheltenham, 2010.
[32] J.P. Holdren, C.R. Sunstein, I.A. Siddiqui, Policy Principles for the U.S. Decision-Making Concerning Regulation and Oversight of Applications of Nanotechnology and Nanomaterials: Memorandum to all Heads of Departments and Agencies, Executive Office of the President, Washington, DC, 2011.
[33] H. Throne-Holst, A. Rip, Complexities of labelling of nano-products on the consumer markets, Eur. J. Law Technol. 2 (3) (2011). Available from: http://ejlt.org//article/view/83
[34] Food and Agricultural Organization and the World Health Organization, Report of the Joint FAO/WHO Conference on Food Standards, FAO/WHO, Geneva, 1962.
[35] Codex Alimentarius Commission, About Codex, 2015. Available from: http://www.codexalimentarius.org/about-codex/en/
[36] Codex Alimentarius Commission, Codex Standards, 2015. Available from: http://www.codexalimentarius.org/standards/en/
[37] R. Mackenzie, The international regulation of modern biotechnology, Yearbook Int. Environ. Law 13 (1) (2003) 97–163.
[38] D.E. Buckingham, P.W. Phillips, Hot potato, hot potato: regulating products of biotechnology by the international community, J. World Trade 35 (1) (2001) 1–32.
[39] Codex Alimentarius Commission, Procedural Manual, twenty third ed., FAO and WHO Food Standards Program, Rome, 2015.
[40] N. Kalaitzandonakes, P.W. Phillips, GM food labeling and the role of the Codex, AgBio Forum 3 (4) (2001) 1.
[41] D. Burke, The dead hand of regulation–who pays the cost of excessive regulation on GM food?, Biologist 51 (2004) 63.
[42] S. Smyth, P.W.B. Phillips, Labeling to manage marketing of GM foods, Trends Biotechnol. 21 (2003) 389–393.
[43] H.I. Miller, D.L. Kershen, A label we don't need, Nat. Biotechnol. 29 (11) (2011) 971–972.
[44] Food and Agricultural Organization and the World Health Organization, Report of the Thirty-Ninth Session of the Codex Committee on Food Labeling, REP11/FL, Appendix III, 2011. Available from: http://www.codex-alimentarius.net/web/archives.jsp
[45] World Trade Organization, Understanding the WTO: Who we are, 2016. Available from: https://www.wto.org/english/thewto_e/whatis_e/who_we_are_e.htm
[46] S.J. Smyth, J. Falck-Zepeda, Socio-economic considerations and international trade agreements, J. Int. Law Trade Pol. 14 (1) (2013) 18–38.
[47] C.G. Gonzalez, Genetically modified organisms and justice: the international environmental justice implications of biotechnology, Georgetown Int. Environ. Law Rev. 19 (2007) 583–642.
[48] Codex Alimentarius Commission, Food Labelling, fifth ed., WHO and FAO, Rome, 2007.
[49] J.C. Cheftel, Food and nutrition labelling in the European Union, Food Chem. 93 (2005) 531–550.
[50] E.F. Einsiedel, Consumers and GM food labels: providing information or sowing confusion, AgBio Forum 3 (4) (2000) 231–235.
[51] A. Weir, P. Westerhoff, L. Fabricius, K. Hristovski, N. von Goetz, Titanium dioxide nanoparticles in food and personal care products, Environ. Sci. Technol. 46 (4) (2012) 2242–2250.

[52] A. Maynard, No, metal oxide nanoparticles in your food won't kill you, The Conversation, 3 June 2014. Available from: http://theconversation.com/no-metal-oxide-nanoparticles-in-your-food-wont-kill-you-27545
[53] J.A. Caswell, Should use of genetically modified organisms be labeled?, AgBio Forum 1 (1) (1998) 22–24.
[54] J.A. Caswell, Labeling policy for GMOs: to each his own?, AgBio Forum 3 (1) (2000) 53–57.
[55] L. Pray, A. Yaktine, Nanotechnology in Food Products: Workshop Summary, Institute of Medicine of the National Academies, Washington, DC, 2009.

CHAPTER TEN

Critical Review of Relevant Recent Patent Applications Related to the Use of Nanotechnology in Food

Peter Viberg
Awapatent AB, Lund, Sweden

Contents

10.1 GENERAL INFORMATION ON PATENTS AND PATENT APPLICATIONS

A patent relates to an exclusive right granted by a state to an inventor with regard to an invention. The scope of protection, and hence the exclusive right, is defined by the patent claims. The right is territorial and applicable in one or more states in which the patent application has been filed and the patent granted. A patent gives its owner a right to exclude others from using the invention, although others may be allowed to use the invention if permitted by the owner.

Although patent law varies between different states, a patent may typically be granted for an invention of technical nature which is novel, involves inventive step, and is industrially applicable. The patent involves an agreement between the state to the owner of the patent in exchange for a detailed description of the invention being published. The right is granted for a limited time period, generally for a period of 20 years from the date of filing of the patent application, during which time maintenance fees must be paid to

Emerging Nanotechnologies in Food Science

keep the patent in force. A patent application typically is published, and thereby publically available, 18 months from the date of filing of the patent application, and eventually the patent is also published after having been granted.

The procedures for obtaining patents in some countries are made more efficient for the applicant by harmonization of patent laws between countries and in some cases the examination of the patent application is common between countries and executed by one single patent office. For example, the European patent office, the EPO, centrally handles one application designating a plurality of European states. Information on patent applications and patents may be obtained from the Patent office of a specific state, and also from the World intellectual property organization, WIPO.

A number of commercial and noncommercial databases are available, in which published patent applications from a large number of states are searchable and accessible. Searches in the databases can be made in the English language, since in many cases at least the abstract of a patent application is available in English, either resulting from the language of filing being English, or from manmade or machine translation. To further facilitate searches in the databases, all inventions are classified based on field of technology by the patent office.

If comparing patent applications to scientific papers, it might be concluded that scientific papers are frequently more elaborate in the experimental sections, although patent applications often contain experimental results to support the scope of the invention where relevant. The journals typically require more technical evidence in the articles compared to patent applications, and the patent applications are not subjected to the peer-review process applied by the journals. The patent claims, which define the scope of protection, often are generalizations of examples disclosed in the application and therefore to some extent broader than experimental examples of the application. Therefore, for example, in a case wherein the invention relates to a composition of compounds, the patent claims typically contain ranges of the content of each compound rather than exact contents. Defining the exact content of the compounds would provide a too narrow scope of protection which could be easy to design around for a competitor.

10.2 PATENTING IN THE FIELD OF NANOTECHNOLOGY AND FOOD TECHNOLOGY

Due to the novelty requirement for inventions, patent applications are often a useful source for obtaining information in the forefront of a technical field, although as mentioned earlier, patent applications will normally not be published until 18 months from the date of the filing of the patent application. Since the almost explosive development of nanotechnology in the 1990s, a very large number of patent applications have been filed related to this field. The interdisciplinary nature of nanotechnology, in combination with the high numbers of patent applications filed, used to make it troublesome to retrieve patent documents in this area from the patent databases. As one result of this, patent offices worldwide started to classify nanotechnology in a class of its own in 2011, the "B82Y" class in the IPC system and later also in the CPC system. Specifically, the "B82Y" class relates to "Specific uses or applications of nanostructures; measurement or analysis of nanostructures; and manufacture of nanostructures". Since long before, different classes have been related to various aspects of food and food technology inventions.

Searches in the patent database Patbase (per December 1, 2015) in the CPC "B82Y" resulted in 61,635 hits, and in classes related to food (CPC "A21," "A22," and "A23") in more than 100,000 hits, thus, clearly illustrating the high patent activity in these fields. Notably, from Figs. 10.1 and 10.2, the trends for patenting in these fields have increased considerably when compared to 30 years ago. Due to the lag period between the filing of a patent application and its publication the last years filings have not been included in Figs. 10.1 and 10.2. Although the "B82Y" class is excellent when searching for patent applications related to nanotechnology, it may be too limited when attempting to search for and review patents concerning use of nanotechnology in food, which in part explains a relatively low number of hits when combining the class "B82Y" with "A21," "A22," or "A23," which resulted in 187 hits, or when using the search term "Food*" in the CPC "B82Y," which resulted in only 464 hits.

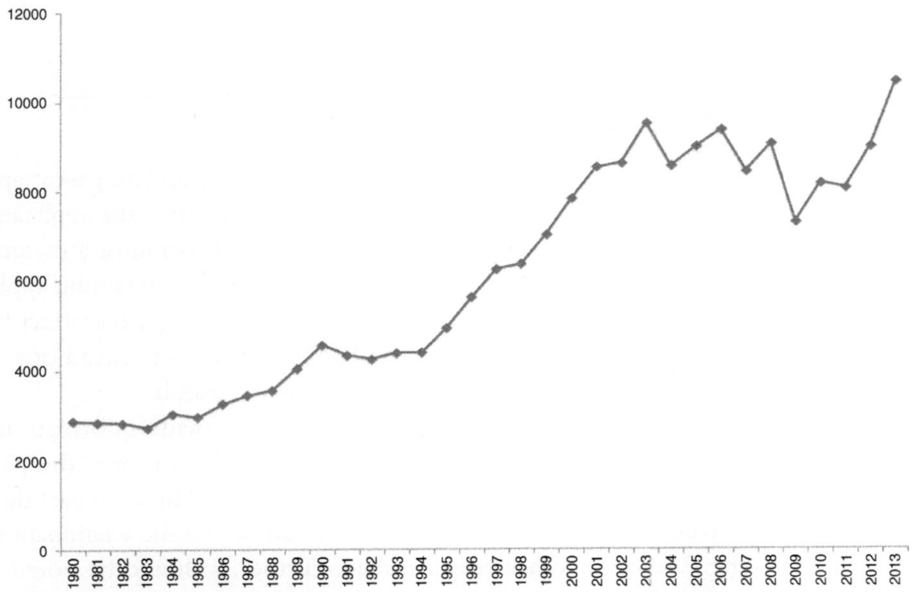

Figure 10.1 ***Number of annually filed patent applications related to food between 1980 and 2013.*** Searches made in database Patbase in classes CPC "A21," "A22," and "A23."

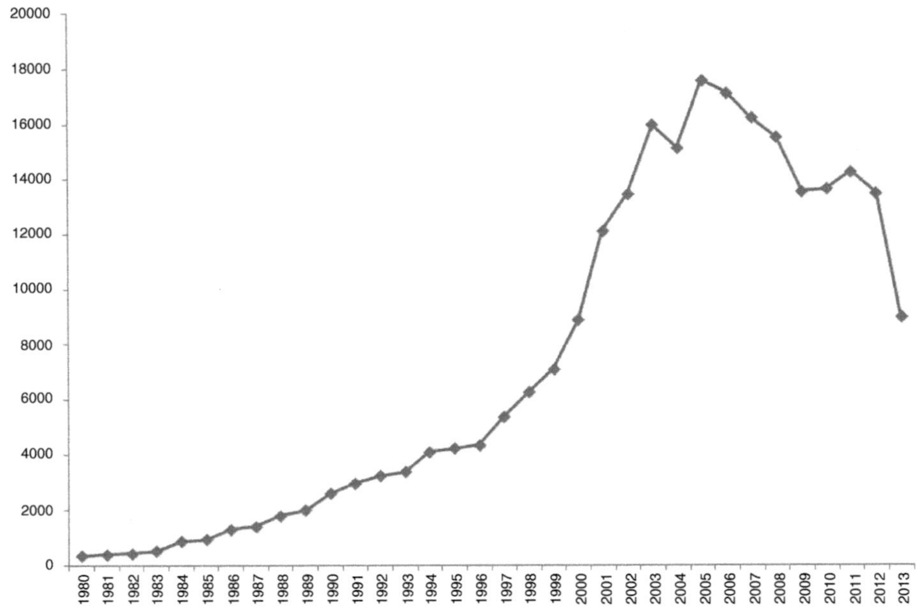

Figure 10.2 ***Number of annually filed patent applications related to Nanotechnology between 1980 and 2013.*** Searches made in database Patbase in class CPC "B82Y."

10.3 FRAMES FOR THE CHAPTER

It has been considered outside the frames of this chapter to translate patent applications which are not available in English. Therefore, analysis of many of the applications herein are based solely on Abstracts, where a legible machine translation or a manmade translation of the abstract has been available in English in spite of the remaining application text being written in non-English language. In many cases an English abstract from an earlier application within the same patent family has been analyzed instead of a corresponding non-English application resulting from the database search.

Note that the present review in no means is intended to provide information on whether a reviewed application has been or will be granted a patent, or if the patent application is maintained or has been abandoned by the applicant. This is in part due to the patenting system where an application is often published before the examination of the application by the Patent office has been finished. Further, it should be noted that among the presented hits from the database search not only patent applications may be found, but also utility models and design applications.

10.4 DATABASE SEARCHES

Due to the high number of patent applications filed in the fields of nanotechnology and food technology, searches using search terms combining food and nanotechnology were conducted in the patent database Patbase to reduce the number of hits. The searches were further restricted to applications published, not filed, during 2014 and 2015. Patbase is one of the several available commercial patent databases. According to their webpage, Patbase covers over 47 million patent families from over 103 issuing authorities.

10.5 REVIEW

Selected and reviewed documents from the searches have been divided into subcategories, which are reflected in the headings, including nanoparticles as food additives or for encapsulation or transporting of compounds; analytical systems and sensors (further divided into subcategories); antibacterial applications; and food packaging technology. Many of the reviewed documents relate in one way or another to nanoparticles with applications in different food related areas.

10.5.1 Nanoparticles for encapsulation and transporting of compounds or used as food additive

Several of the reviewed documents relate to nanoparticles for encapsulation or transporting of compounds with applications in the field of food, as well as nanoparticles used as food additives, that is, food additives in the form of nanoparticles. This area has been addressed in Chapter 2 of this volume.

Some of the documents relating to nanoparticles for encapsulation and transporting of compounds disclose nanoparticles based on proteins. Nanoparticles having a casein matrix is reported to enable encapsulation of water-soluble or fat-soluble biologically active compounds with application in the food sector [1]. More specifically, the nanoparticles in addition to casein contain a basic amino acid, for example, arginine or lysine, and a metal suitable for food, such as calcium. It is reported that the nanoparticles are stable and capable of protecting biologically active compounds from degradation by external agents, such as light, pH changes, and oxidation, both during product processing and storage. When applied in food, the nanoparticles are reported to protect biologically active compounds from the acidic conditions of the stomach, thus preventing release of the compounds along the gastric tract, and avoiding their precipitation and resulting reduced bioavailability. Nanoparticles having a zein matrix (i.e., a type of plant protein matrix) and a basic amino acid, with applications in the food sector, are disclosed in one document [2]. These nanoparticles also function to enable encapsulation of water-soluble or fat-soluble biologically active compounds. Another application [3] relates to the field of food technology and delivery of hydrophobic biologically active compounds, particularly nutrients, via food products and beverages using isolated casein micelles. A method for the production of capsules, based on milk serum proteins, including electro-drawing or electro-spraying or blow-drawing or blow-spraying has also been described [4]. It is disclosed that the capsules can be used as vehicles for the encapsulation of functional additives and ingredients for the incorporation in pharmaceutical or food preparations.

Nanocapsules having low oxygen permeability and thus protecting liquids included in the core of the nanocapsules from oxygen, specifically intended for protection of food or food ingredients from oxidation have been described [5]. For this purpose, the nanocapsules have a specific design: they comprise a shell and a core, wherein the core contains at least one compound characterized by being a liquid between 0 and 80°C, and the shell contains semicrystalline cellulose nanofibers and/or semicrystalline cellulose nanocrystals each having a crystallinity index of at least 30%. Another invention [6] relates to monodisperse particles having sizes between 100 and 400 nm prepared from natural waxes from the Amazon. The nanoparticles are reported to allow for the encapsulation of active agents and pharmaceuticals and applied in the development of food components. Nanoparticles made of sugar-beet pectin, and a bioactive compound such as a nutraceutical or a drug bound to the sugar-beet pectin, useful for foods, beverages, or pharmaceutical preparations are disclosed in one document [7].

Nanoparticles specifically used as food additives, rather than for carrying or encapsulating compounds, will now be discussed. Two documents [8,9] disclose preparation of a nanoparticulate or micro-particulate protein-based products to be used as fat replacer in food products. One of the preparation methods [8] comprises the steps of mixing a polysaccharide solution, a protein solution, and an alkaline earth metal salt solution. The mixture is heated to form a suspension of alkaline earth metal cation-stabilized protein microparticles or nanoparticles, followed by rapid cooling of the mixture. One specifically mentioned example includes whey protein and konjac gum. A nanogel composition comprising at least one water-soluble active ingredient, one or more plant proteins, and one or more soy-soluble polysaccharides is disclosed in another document [10]. The compositions, it is suggested, can be used for the enrichment, and fortification of food, beverages, animal feed, and cosmetics further allowing stabilization of the active ingredient.

Calcium hydrogen phosphate as nanoparticles used as food additive has been disclosed [11]. The production of the nanoparticles is based on an emulsion method, wherein calcium hydroxide and phosphoric acid each form a reverse emulsion in an organic phase. The two emulsions are mixed and stirred uniformly, after which a stabilizer is added. The, thus, obtained product is reported to have small and uniform particle size and good fluidity, and to allow for efficient absorption by the human body as well as an improved utilization rate of both phosphorus and calcium.

Two reviewed inventions [12,13] relate to minimizing salt ingestion and low sodium salts by taking advantage of small particles of salt. One of these inventions [12] relates to particles of salt having average diameters of 180–350 nm obtained by mixing a solution of table salt and a carrier in a weight ratio of 1.8–2.2:0.8–1.2. The solution is then spray dried and pulverized into the particles. It is reported that the salty taste is improved with these salt particles as compared with common salt. The other invention [13] discloses small salt particles adhered to an edible carrier particle. The increased surface area-to-volume ratio of the product is reported to assist in interacting with mouth physiology to provide increased salt flavor as compared to conventional table salt, while substantially reducing the amount of salt ingested by the consumer.

10.5.2 Analytical systems and sensors

One large field in the combined area of nanotechnology and food includes different types of analytical systems and sensors for food analysis, often with an emphasis on food safety and health. In an attempt to improve the clarity, the reviewed documents have been divided into subcategories related to *Detection of microorganisms*, *Detection of human or animal pharmaceuticals in food*, *Detection of nutritional contents in food*, and *Detection of other compounds in food*. Different types of binding sites of sensors are disclosed in the reviewed documents, including aptamers, antibodies or immunoassays, and molecular imprints. Magnetic nanoparticles used for extracting molecules of interest from samples are used in several of the aforementioned applications.

10.5.2.1 Detection of microorganisms

A major part of the reviewed documents concerning detection of microorganisms relies on biosensors including antibodies or aptamers. It will be noted that a vast majority of these reviewed inventions relate to detection of food borne pathogens.

Aptamers, that is, oligonucleotide or peptide molecules binding to a specific target molecule, have been used for selective binding of microorganisms in two of the reviewed applications [14,15]. A biosensor chip intended for rapid detection of *Salmonella typhimurium* is suggested. The chip is obtained by attachment of aptamers with a specific gene sequence on a chip modified with gold nanorods. It is reported that the *S. typhimurium* can be selectively identified even in a complex medium. Further, from the measurement of the local surface plasma resonant wavelength of the chip before and after contacting with *S. typhimurium*, quantitative detection is achieved [14]. The presence or absence of microorganisms in a sample has also been detected by using two different aptamers: a first aptamer binding to a cell-surface protein of the microorganism under formation of a complex, and a second aptamer binding to the first cell-surface protein or a second cell-surface protein of the microorganism. An assay is carried out to detect the second of these aptamers. The detection of the second aptamer indicates that the microorganism is present in the sample, and vice versa. [15]

A high number of documents relate to the use of antibodies for binding of target microorganisms. One application [16] takes advantage of titanium dioxide nanowires for detecting pathogenic bacteria of food. Antibodies are adhered to the surface of modified nanowires having diameters of 60 nm through covalent bonding. The thereby obtained nanobiosensors are reported to benefit from rapid monitoring of multiple bacterial pathogens and chemical substances. Rapid detection of the food-borne pathogenic bacteria *Enterobacter sakazakii* has also been reported [17]. The disclosed method comprises adding immunized superparamagnetism nanobeads to a sample, followed by a stabilizer, and incubation whereby the magnetic beads are specifically bonded and enriched on *E. sakazakii*.

Two inventions disclose detection methods for food-borne pathogenic bacteria [18,19]. Both are taking advantage of immunomagnetic Fe_3O_4 nanomaterials. According to one of the methods [18] the Fe_3O_4 nanomaterial is used for preparing an immunomagnetic bead for specific enrichment of a target bacterium, and the influence of paramagnetic and superparamagnetic characteristics of Fe_3O_4 on nuclear magnetic resonance relaxation time is used for determining any presence of the bacterium. It is reported that the method enables rapid detection of harmful pathogenic bacteria in a food sample thereby allowing rapid screening of large quantities of samples. According to the other method [19], detection of pathogens in a sample is carried out by separating the nanomagnetic beads once having captured the target bacteria, wherein iron ions are converted through nitration reaction. Quantification of target bacteria is indirectly realized via quantification of the iron ions.

Detecting and identifying specific microorganisms in a culture sample has been targeted [20]. Raman spectroscopy active nanoparticles with binding members with affinity for the microorganisms of interest are used for forming a complex of nanoparticles and microorganisms in the culture sample. Antibodies are given as examples of the binding members. Further, magnetic capture particles, also having one or more binding members with an affinity for the microorganisms of interest, are suggested to be used to capture and concentrate the complex antibody-microorganism for detection and identification.

One invention [21] relates to an electrochemical sensor for detecting *Escherichia coli*. The sensor comprises an electrode, a prussian blue-carbon nanotube-nanogold compound layer, an *E. coli* antibody layer, and a bovine serum albumin layer. The electrochemical sensor is reported as being suitable for detection of the *E. coli* for food safety purposes.

Three separate applications [22–24] belonging to different patent families but from the same inventors relate to rapid detection of food borne pathogenic bacteria using Fe_2O_3 containing nanoparticles. The inventions take advantage of the specificity of an antibody I being combined with the target bacteria. The Fe_2O_3 containing nanoparticles are utilized to prepare immune-magnetic beads of an antibody II to enrich target strains marked by the antibody I. The nanoparticles bound to the target bacteria are captured through separation, and then subjected to nitration to be converted into ferrous ions capable of being detected and thereby, indirectly, detecting whether the sample contains the target pathogen. Under certain conditions and within certain ranges the detection is reported to allow quantitative detection of the target bacteria.

As indicated from the abstracts of the documents, one of the documents [22] relates to gamma-Fe_2O_3@Au composite nanoparticles, another of the documents [23] relates to gamma-Fe_2O_3 nanoparticles, and the third document [24] relates to Fe_3O_4 and Au-nanomaterial. It has not been further concluded from the abstracts how or if the Fe_2O_3 comprising nanoparticles of the three documents differ from each other.

Some of the reviewed documents did not clearly specify, at least not in the abstract, the type of binding site being used. One such document [25] relates to a method and apparatus for separating and detecting food poisoning bacteria using magnetic nanoparticles. The nanoparticles, which are described as having an affinity for the bacteria, are contacted with a sample, thereby allowing separation and analysis of the bacteria from the sample. A different approach [26] discloses a membrane strip biosensor device which may be used to detect pathogens, proteins, and other biological materials of interest in food, water, and environmental samples. The device uses a fluid mobile conductive composition of ferromagnetic particles bound to a conductive polymer in turn bound to a capture reagent. Another method [27], based on NMR, for the detection of food-borne pathogens is based on the use of Fe_3O_4@Au composite nanoparticles. The detection of target pathogens in the sample is carried out through the influence of paramagnetic and superparamagnetic characteristics of the nanoparticles on the NMR attenuation signal relaxation time. It is discussed that the method can be used for quick screening of a large number of samples. A method for detecting pathogenic bacteria in food by means of magnetic beads designed to be suitable for being added into a PCR system has also been reported [28]. DNA in food samples is captured by the magnetic nanoparticles, which are further processed, and eventually PCR detection is directly carried out on the nanoparticles. The method is reported to have advantages including being rapid, high in sensitivity, and capable of detecting multiple kinds of pathogenic bacteria simultaneously. A method for detecting food poisoning bacteria, including a method for rapidly and quantitatively isolating the bacteria from food [29] involves introducing magnetic nanoparticles to a sample for binding to the bacteria. The nanoparticles are thereafter isolated and transported through a solution having high viscosity so as to separate the nanoparticles to which bacteria are bound from nanoparticles to which no bacteria are bound, the former are then quantified.

10.5.2.2 Detection of human and animal pharmaceuticals in food

A field attracting interest, not only in media and research, but also in the patenting field is detecting undesirable presence of pharmaceuticals and particularly antibiotics in food. For example, dairy products could potentially contain antibiotics originating from treatment of the cows. A number of relevant inventions will be described to give an overview of trends in this field.

One method for detecting veterinary medicine residues in foods using magnetic nanoparticles with antibodies for binding of the veterinary medicines has been disclosed [30]. The method takes advantage of that the magnetic susceptibility of the nanoparticles is being reduced when mixed with a sample containing the veterinary medicine. Thereby, quantification of the amount of the veterinary medicine in the food is allowed. Another method [31] similarly relates to detecting animal drug residues in food using bio-functionalized magnetic nanoparticles as labeling antibodies for the drug. Again, reduction in magnetic susceptibility of the nanoparticles resulting from the association between the nanoparticles and animal drugs is used for quantifying the drug residues in food.

One method for detection of antibiotics in liquid milk products takes advantage of magnetic particles used in a solid-phase extraction method for enriching beta-lactam penicillin antibiotics in the product [32]. The method involves adding superparamagnetic nanometer or submicrometer magnetic particles having surfaces functionalized with aniline to milk, after which the magnetic particles are separated using magnetism, whereby a liquid enriched with the antibiotics is obtained. Another method relates to the detection of fluoroquinolone drugs in milk (fluoroquinolone is a type of broad-spectrum antibiotics) [33]. A hollow imprinted material with ofloxacin as a template was synthesized and used as an adsorbent. High performance liquid chromatography with UV detection was used for detection of the fluoroquinolone drugs.

Some documents [34,35] relate to aptamer sensors for detecting oxytetracycline, also a type of broad-spectrum antibiotic, in food. Another concerns analysis of the antibiotic doxycycline using a nano-SiO_2-based doxycycline molecularly imprinted polymer [36]. The molecularly imprinted polymer was obtained by using nano-SiO_2 as a support material/carrier and doxycycline as a template molecule. After polymerization, SiO_2 was removed by etching with hydrofluoric acid leaving an imprinted polymer. The imprinted polymer was used as a packing material in a solid phase extraction column and detection of tetracycline antibiotics in animal derived foods.

An electrochemical sensor for detecting clenbuterol hydrochloride (a substance used for treatment of breathing disorders, but also known to be used as a doping agent for humans and animals) in food analysis has been described [37]. The sensor integrates three working electrodes, and enables simultaneous detection of hydrochloride of ractopamine, salbutamol, and clenbuterol in a range of 0.01–1000 ppb. One invention [38] discloses a sibutramine magnetic molecularly imprinted polymer for use in detecting sibutramine present in food. Magnetic molecularly imprinted polymer particles composed of a core made of Fe_3O_4@SiO_2 magnetic nanoparticles and a shell made of sibutramine molecularly imprinted polymers were used. The nanoparticles are controlled by an external magnetic field, thereby facilitating efficient absorption, washing, and desorption.

10.5.2.3 Detection of nutritional contents in food

Inventions related to analysis of antioxidant and sugar levels of food are discussed subsequently; these constitute an important but smaller part of the overall number of inventions in the food nanotechnology area.

A method and an assay for portable colorimetric detection of antioxidants in food samples comprising immobilized cerium [IV] oxide nanoparticles has been described. The method involves the contact of the nanoparticles with a food sample containing antioxidants, and an optical property of the colorimetric reagent associated with the presence of antioxidant in the food sample was detected and quantification of the antioxidant was achieved [39].

One invention relates to method for detection of antioxidant capacity of plants and plant-derived liquid food being based on a system of gold nanoparticles [40]. The method takes advantage of a correlation between optical properties of nanogold particles and the antioxidant capacity of a sample. The system of gold nanoparticles contains $AuCl_4$, CTAB, sodium citrate, and a phosphate buffer. The sample to be analyzed is added to the system, followed by heating, whereby $AuCl_4$ is reduced to gold by the antioxidant activity of the sample resulting in shifting the color of the system to wine red. The color shift allows determination of the antioxidant capacity of the sample.

Some reviewed inventions relate to determination of sugar levels in food or blood [41–43]. One of the inventions [41] relates to measuring sugar through gold nanoparticle colorimetry. Phenyl boronic acid and gold nanoparticles are added to a sugar-containing sample. The combination of the sugar and the phenyl boronic acid inhibits combination of the phenyl boronic acid and the nanometer gold, which results in a color shift of the solution, thus enabling quantitative or qualitative detection of the sugar in the sample.

Another method relates to determining glucose using gold nanoclusters as a fluorescence probe [42]. The method takes advantage of glucose being oxidized by glucose oxidase to generate H_2O_2. Fe^{2+} in turn catalyzes the H_2O_2 to generate hydroxyl radicals which quench fluorescent light of the gold nanocluster. Thereby a change of a fluorescence emission spectrum is obtained which allows determining the glucose content. Another method for determining glucose is based on a hollow copper sulfide/polypyrrole nanometer sized compound having activity similar to peroxidase, and which can generate hydroxyl radicals with strong oxidizing ability by catalyzing H_2O_2. The invention is claimed to be applicable to detection and analysis of glucose content in food and blood [43].

Measuring of salt content in drinking water has been realized by a method using gold nanoparticles having different colors in different salinities [44]. The method comprises preparing and modifying the gold nanoparticles, and, testing a saline solution with different salt concentrations. A test paper is described which allows for determining the salinity of a sample by comparing with a colorimetric card.

10.5.2.4 Detection of other compounds in food

Now the focus shifts to analysis of compounds other than bacteria, drugs, and nutrients.

Acrylamide is one compound which may form in some types of food during cooking and which may, potentially, increase the risk of cancer. Analysis of acrylamide has gained interest not only in academic research but also in the patent field. Detection of acrylamide in thermally processed foods is addressed in one invention [45], which relates to a method comprising binding of acrylamide to nanosized gold particles, and quantifying by means of UV-visible absorption spectrophotometry. Another method for detecting acrylamide takes advantage of a sol-gel molecular imprinted electrochemical sensor based on nanomaterial composites [46]. The method comprises coating a composition of carbon nanotubes, gold nanoparticles, and chitosan on the surface of a glassy carbon electrode, and electrochemical deposition in a sol-gel solution containing template molecules, functional monomers, and cross-linking agents. After removal of the template molecules acrylamide was detected in the sample by using the, thus, obtained molecular imprinting electrochemical sensor. It is disclosed that the method provides sensitivity that enables detection of acrylamide in fried foods such as potato chips.

Detection of pesticide residues in processed food and agricultural products using DNA labeled fluorescent magnetic nanoparticles has been described [47]. DNA was grafted on the particles, which were intended to be applied in an antigen–antibody reaction in a quantitative analysis method.

Some reviewed inventions relate to detection of melamin in foodstuffs [48–50]. Notably, it has been reported in media that melamine, being a toxic organic chemical rich in nitrogen, has been used to increase the nitrogen content of diluted and adulterated milk, in attempts to increase the apparent protein content of diluted milk and thereby to mask the dilution. Addition of melamine to food is not approved by the FAO/WHO Codex Alimentarius (food standard commission) [51]. Thus, there is a need for detection of melamin in, for example, dairy products. One invention [48] dealing with this discloses an electrochemical method for detecting melamine in dairy products or food. A composite material consisting of carbon nanotubes and natural polymers being coated on a glassy carbon electrode was used as a sensor. Electrostatic attraction and hydrogen-bond interaction between the composite material and the melamine was followed by detection of melamine in dairy products or food using a tri-electrode system. Other inventions [49,50] also relate to methods for detecting melamine in food. These inventions take advantage of gold nanoparticles being modified with nucleic acid aptamers, and the binding of melamine to these modified nanoparticles. One of the documents claims to enable a detection limit 4.3 ppb, which is of high sensitivity in relation to the current state-of-the-art.

Other inventions relate to detection of pigments or dyes in food. For example, a method for detecting synthetic pigments in food by a Raman spectrometric method has been described. Nanostructured gold modified by mercaptoethylamine was used in synthesizing nanostructured gold sol with positive charge. The nanostructured gold sol then served as a Raman spectrum enhanced reagent for detection of the pigment. Absorption of the pigments occurred through electrostatic interactions [52]. Another method relates to analysis of trace amount of rhodamine 6G, a fluorescent dye, in food samples [53]. The method uses a nanoscaled rhodamine 6G molecularly imprinted solid phase extraction material. Surface-aminated Fe_3O_4 nanoparticles functioned as a magnetic core having a layer of rhodamine 6G molecularly imprinted polymer. It was reported that the material was used for selective adsorption, extraction, and separation of trace amounts of rhodamine 6G in food sample.

10.5.3 Antibacterial applications

Above under Analytical systems and sensors, a large section was committed to detection of microorganisms by means of sensors having nanoscaled structures. Another common use of nanostructures relates to their antibacterial properties, which is also reflected in patent documents. Many of those documents relate to such antibacterial applications from the presence of nanostructures in materials and packages intended for food. Particularly nanoscale silver structures are reported in a plurality of the documents [54–57]. The antimicrobial properties of silver have been recognized for a long time, although its use in food packaging may be regarded as being controversial or challenging due to potential health risks. Chapter 6 has compiled the toxicity of silver nanoparticles, among other types of nanoparticles, with a wide range of bacteria. Negative health effects of nanoparticles in humans are also of concern. Some of the patent applications attempt to tackle these problems.

One invention [54] relates to an absorbing pad having antimicrobial properties derived from antimicrobial silver nanotechnology. The pad is intended to be placed at the bottom of plastic food containers. Another invention [55] discloses nanostructured silver solutions embedded in ice for preservation of foods including fish, chicken, and mutton. A nanofiber material for pork preservation has also been described [56]. This material is prepared from catechin (an antioxidant), copper nitrate, and polyvinylpyrrolidone as raw materials. By adopting the nanofiber material, a problem relating to easy oxidation of catechin in antibacterial applications is claimed to be solved, thus resulting in improved antibacterial properties of the nanomaterial. The nanofiber is claimed to be nontoxic. A nanometer sized silk fibroin-nanosilver antibacterial composite membrane has also been described [57]. The membrane is reported to have potential use for, for example, food packages, hygienic products, and medical materials. A package specifically intended to hold foodstuff susceptible to microbial growth is addressed with another invention. The interior of the package has an antimicrobial surface with an ordered nanoarray of metal or metal oxide nanostructures. Silver oxide is provided as an example. The nanoarray is claimed to reduce the amount of silver required to achieve an antimicrobial effect, to less than 0.001% by weight, as compared to convention technology. Further, due to the nanostructures being rigidly anchored to the package surface risks of ingestion of the nanostructures by consumers are reported to be minimized [58].

Dendrimer nanoparticles conjugated with vancomycin and/or polymixin (types of antibiotics) have been used to sequester and identify bacteria and to screen liquid samples including water samples, and food samples for the presence of bacteria [59].

10.5.4 Food packaging technology

Packaging materials including nanotechnology have been discussed in depth in Chapters 7 and 8. A brief overview of relevant inventions will be provided here.

Two inventions [60,61] relate to packaging materials containing soy bean proteins. One of the inventions [60] relates to a nanocomposite soy protein plastic and its preparation. The plastic is reported to have high water resistance and to benefit from being environmentally degradable. The other invention [61] relates to a soybean protein isolate-polyaniline antimicrobial composite film. The material for the film is synthesized from soybean protein isolate and polyaniline complex. The polyaniline is in the form of nanofibers having a diameter of 40–45 nm.

Several reviewed inventions relate to packages or packaging films intended to provide air or oxygen sealed environments for prevention of oxidation of food. A food container made of a plastic material with a nanostructured hydrophobic surface has been described [62]. The container is reported to provide efficient gas blocking as well as high hydrophobicity. A composite film made of a modified nanocellulose obtained from Rami, that is, a plant in the *Urticaceae* family, and polylactic acid having high strength as well as good properties relating to air permeability, biodegradability, stability, softness, and biocompatibility is disclosed in one application [63]. Another invention [64] relates to a composition of solid lipid nanoparticles used as a nanocoating for natural fresh foodstuffs. The composition contains solid lipids or wax, emulsifying stabilizing agents, and film-forming materials in an aqueous dispersion or solution. The foodstuffs are coated by means of fluidization, immersion, or spraying of the composition.

An edible membrane based on a composition of *Thamnaconus modestus* fish skin nanoparticles has been invented [65]. The membrane is intended for being used for food or medicine packing, and for preventing lipids from being peroxided. A method of providing a package with an oxygen barrier for pourable food products is disclosed in one document [66]. The liquid oxygen barrier composition of a polymer dispersion containing nanoparticles of clay or silica is applied as a coating on the outside of packages. A further oxygen barrier film is disclosed in another document [67], which uses graphene oxide nanoribbons in the film. The film is reported to have excellent blocking performance and good acid and alkali resistance. Ice bags for packaging ice and frozen products, and food packaging films are included among the suggested uses of the film.

Further use of silver in antibacterial materials has been realized by a silver nanoparticle/lignin composite material [68], wherein the silver nanoparticles are adsorbed to the surface of lignin. The reduction of *E. coli* is reported to be greater than 75% after 2 h.

A plastic material with antibacterial properties based on mixing nanoscale metal ion material with plastic particles followed by injection molding arranged on the container cover has been manufactured [69].

Besides packaging materials with antibacterial properties, degradable packaging materials also have great interest and several novel applications already exist, for instance a degradable chlorinated polypropylene/SiO_2 nanocomposite membrane which is reported to be completely degraded within 1–2 years [70]. Suggested uses of the degradable membrane include preparing of food preservative films, degradable foam material, disposable tablewares, disposable medical materials, and a food packaging material.

10.6 DISCUSSION AND CONCLUSIONS

Patenting over the last 30 years has increased considerably in the field of food, and gone from essentially nothing to a high number of new applications per year in the field of nanotechnology. The combined field of nanotechnology in food naturally involves lower total numbers of patent applications, although the development in that field, speculatively, also has accelerated.

From the reviewed patent applications, certain focus areas are evident: the use of nanoparticles for protection and transport of compounds added to food; different type of food analysis systems and methods taking advantage of nanoscaled structures; antibacterial applications, such as for food preservation; as well as the use of nanoscaled structures in food packages. Of those focus areas, food analysis has attracted most attention among the reviewed documents.

Common to all of the focus areas is a health aspect to which a clear majority of the reviewed documents relate. Many of the documents tackle detection or reduction of toxic, unhealthy or otherwise undesirable microorganisms or compounds in food. Other documents disclose methods and means for preservation of food, which also relates to health.

When discussing trends and activity in patenting in a field of technology, it is interesting to consider some aspects of patenting and research. In high technology fields, such as nanotechnology applied to food, patenting is frequently based on considerable research activity. Therefore, at least to some extent, trends in patenting may be expected to follow paths and trends of research in that field. However, while much relevant research is followed by publications in journals, certainly not all relevant research, from academia or industry, results in the filing of patent applications. Although the reasons may vary, there is a financial aspect: the patenting process involves costs which should be motivated by gains resulting from the exclusive right to the invention obtained from the patent. Thus, trends and activity in patenting are expected to depend also on commercial interests in the inventions. Further, there may be various reasons not to publish research by filing of patent applications. Therefore, trends and activity in patenting may reflect the research in a different way than scientific papers do, but are likely relevant indicators of technological development and commercial interest in that particular field of technology.

Good health is, naturally, central to man, which not only attracts focus in research but also fuels commercial interest. Patenting in the field of food, being so closely related to health, may therefore be expected to have a focus on health now as well as in the future.

Note: Although the database search for the review was limited to applications published during 2014 and 2015, referenced documents as listed below may have earlier publication years, for example where the review was based on an application belonging to the same patent family as the application resulting from the search but having an earlier publication year. The below reference list discloses Inventor[s], Title of invention, Year of publication, and publication number.

REFERENCES

[1] M. Agüeros Bazo, I. Esparza Catalán, C. González Ferrero, C.J. González Navarro, J.M. Irache Garreta, A. Romo Hualde, Nanoparticles for the encapsulation of compounds, preparation thereof and use of same, US2013209530AA (2013).

[2] M. Agüeros Bazo, I. Esparza Catalán, C. González Ferrero, C.J. González Navarro, J.M. Irache Garreta, A. Romo Hualde, Nanoparticles for encapsulation of compounds, the production and uses thereof, US2013116261A (2013).

[3] Y.D. Livney, D.G. Dalgleish, Casein micelles for nanoencapsulation of hydrophobic compounds, US2009311329A (2009).

[4] A. López Rubio, J.M. Langaron Cabello, Method for the production of micro-, sub-micro- and nanocapsules, based on milk serum proteins, EP2724775 (2015).

[5] K. Landfester, S. Balouchev, J. Risbo, A.J. Hanner, Capsule, namely nanocapsule, microcapsule or macrocapsule, having a very low oxygen permeability, WO15059179A1 (2015).

[6] B. Giehl Zanetti Ramos, Nanoparticulas preparadas a base de ceras naturais da amazonia. Processo de obtençao, composiçao e aplicaçoes das nanoparticulas, BRPI0801545A2 (2010).

[7] Y.D. Livney, Pectin based nanoparticles, WO15087329A1 (2015).

[8] M. Auty, A. Abhyankar, L. Wang, A process for preparing a nano-particulate or micro-particulate protein-based product suitable for use as a fat replacer in food products, WO14140023A1 (2014).

[9] L.J.R. Bovetto, C.J.E. Schmitt, M. Beaulieu, N. Carlier, G. Unterhaslberger, Nanoparticulated whey proteins, US2007231453A (2007).

[10] X. Ding, O. Vidoni, P. Yao, Nanogel comprising water-soluble active ingredients, (2015) US2015366243AA (2015).

[11] C. Su, W. Wei, Z. Zhenang, Method for producing nanoscale calcium hydrogen phosphate, CN103771381A (2014).

[12] M.J. Choi, H.Y. Cho, I.H. Jung, B.S. Kim, Preparing method of low-sodium salt comprising nanoparticle and method for preparation thereof, KR20150021254A (2015).

[13] Y.-J. Wang, S.K. Patil, Low-sodium salt compositions, US8900650 (2014).

[14] Chen Zhitao, Biosensor chip for rapidly detecting *Salmonella typhimurium*, CN104749379A (2015).

[15] U.A. Ochsner, N. Janjic, Compositions and methods for detecting microorganisms, WO15126769A1 (2015).

[16] S. Jinming, D. Wenjun, Preparation method of super-long titanium dioxide nanowires for detecting pathogenic bacteria of food, CN104237515A (2014).

[17] Y. Zhao, Y. Chen, H. Huang, L. Guo, Method for rapid detection of *Enterobacter sakazakii*, CN102323408B (2014).

[18] Z. Jinsheng, T. Qun, L. Weihua, NMR detection method for food-borne pathogenic bacteria based on Fe_3O_4 nano-material, CN103207198 (2015).

[19] Z. Jinsheng, P. Hong, L. Weihua, Quick detection method for food-borne pathogens based on immunomagnetic separation of Fe_3O_4 nano materials, CN103116026B (2015).

[20] K. Weidemaier, R.L. Campbell, E.G. Carruthers, A.C. Curry, K.G. Dolan, A. Liebmann-Vinson, W.D. Woodley, M.M.H. Kuroda, A.D. Lentz, D. Livingston, M.J. Lizzi, A.R. Lockhart, E. Ritchey, E.A. Fallows, D.E. Gorelick, J. Kessler, S. Lovette, J.S. Ojala, M.A. Talmer, M. Bartkowiak, S.N. Danhof, G.S. Kramer, T.D. Haubert, M.L. Marshall, J.A. Prescott, R.J. Somerville, M.S. Ulrich, D.S. Sebba, Methods, systems, and devices for detecting and identifying microorganisms in microbiological culture samples, US2015118688A (2015).

[21] H. Jiadong, G. Yu Na, W. Yu, L. Su, C. Min, L. Jie, X. Wei, W. Hongzhi, X. Ying, C. Jie, Q. Tingting, Electrochemical sensor for detecting *Escherichia coli*, CN204203160U (2015).

[22] Z. Jinsheng, T. Qun, L. Weihua, Food-borne pathogenic bacteria quick detection method based on gamma-Fe2O3@Au nano particle indirect enrichment and immunomagnetic separation, CN103185796 (2015).

[23] Z. Jinsheng, T. Qun, L. Weihua, Gamma-Fe_2O_3 nanoparticle indirect enrichment immunomagnetic separation based method for rapid detection of food-borne pathogenic bacteria, CN103217529 (2015).

[24] Z. Jinsheng, T. Qun, L. Weihua, Quick detection method for food-borne pathogenic bacteria based on immune magnetic separation of Fe_3O_4 and Au nano-material, CN103091465 (2015).

[25] S.M. Jeon, D.H. Kwon, Separation method of food poisoning bacteria from large volume sample using magnetic nanoparticle, flow and magnetic capture device, KR20140146845A (2014).

[26] E.C. Alocilja, S. Pal, E.B. Setterington, Electrically-active ferromagnetic particle conductimetric biosensor test kit, US2008314766 (2015).

[27] Z. Jinsheng, T. Qun, L. Weihua, NMR [nuclear magnetic resonance] quick detection method for food-borne pathogens based on Fe_3O_4@Au composite nano particles, CN103115934A (2013).

[28] S. Xianming, B. Yalong, S. Chunlei, S. Minghui, W. Dapeng, C. Yan, Method for rapidly detecting pathogenic bacteria in food, (2014) CN103160577 (2015).

[29] S. Jeon, D.-H. Kwon, J.M. Joo, K.H. Park, S.-B. Han, H.-J. Lee, K.-H. Kim, Method for detecting food poisoning bacteria using magnetic nanoparticles and solution having high viscosity, US2015293083AA (2015).

[30] Yang Xiele, Method for detecting veterinary medicine residues in foods, CN103926399A (2014).

[31] H. Chia Shin, Y. Shieh Yueh, Method and reagent for detecting animal drugs residues in food, TW201120445A (2011).

[32] X. Yonghua, G. Liang, W. Kesheng, Method for enriching beta-lactam penicillins antibiotics in liquid milk, CN103105446A (2013).

[33] Z. Chunhong, T. Wanjin, Z. Haixia, Method for detecting fluoroquinolone drugs in milk, CN104698095A (2015).

[34] X. Wei, H. Jiadong, L. Su, Preparation method and application of competitive type aptamer sensor for detecting oxytetracycline, CN103616423A (2014).

[35] L. Chunxia, T. Zonggui, L. Changbin, S. Fengxia, K. Lichao, C. Xia, W. Guohong, Method for detecting oxytetracycline [otc] in food by virtue of functional magnetic bead isolation-enzyme linked aptamer, CN104634754A (2015).

[36] T. Huarong, S. Xiongyuan, P. Chao, L. Yue, Preparation and application of nano silica-based doxycycline molecularly imprinted polymer, CN103554363A (2014).

[37] W. Qin, Y. Haiqin, W. Huan, Z. Yong, L. He, D. Bin, W. Dan, M. Hongmin, Preparation method of electrochemical sensor for detecting three clenbuterol hydrochloride and application, CN103163193B (2014).

[38] X. Guanhong, D. Peng, Z. Ting, C. Yun, L. Ting, L. Fei, H. Qin, Sibutramine magnetic molecularly imprinted polymer and preparation method thereof, CN103819632A (2014).

[39] E. Sharpe, E.S. Andreescu, D. Andreescu, Portable nanoparticle based assay for rapid detection of food antioxidants [nanocerac], WO14031633A1 (2014).

[40] C. Jia, S. Li, D. Wei, Y. Ye, Method for detecting antioxidant capacity of plant and plant-derived liquid food based on nanogold system, CN101762559A (2010).

[41] H. He, J. Ling, L. Yanping, T. Danbi, L. Shuang, Method for measuring sugar through nanometer gold colorimetry, CN102854181A (2013).

[42] C. Wei, D. Haohua, H. Dong, X. Yanyu, S. Shijie, Method for determining glucose by using gold nanocluster as fluorescence probe, CN104330393A (2015).

[43] Z. Chunling, L. Zibin, Z. Hongmei, X. Zenghong, Hollow copper sulphide/polypyrrole nanometer compound and application thereof, CN104004186A (2014).

[44] W. Ping, H. Dannong, Method for testing low-concentration salinity, CN103674944A (2014).
[45] L.Yanbin, H. Qinqin, F.Yingchun, X. Xiahong, Z.Yi, Method for quickly detecting acrylamide in heat-processed food by use of nano-gold colorimetric biosensor, CN105115968A (2015).
[46] L. Xia, M. Lugang, Y. Yang, Method for detecting acrylamide through sol-gel molecular imprinting electrochemical sensor based on nanomaterial composites, CN104181216A (2014).
[47] S.W. Kim, H.U. Lee, C.H. Park, Y.S. Song, J.H. Lee, D.U. Jung, DNA labeled core/shell nanoparticle having fluorescence, and method of detecting a bioactive material, KR20140094035A (2014).
[48] Z. Tingkai, L. Tiehu, L. Lehao, Electrochemical method for detecting melamine in dairy products or food, CN103439390 (2015).
[49] S. Chunyan, Z. Hongru, S. Fei, G. Jiajia, L. Yeli, X. Jingyue, L. Ying, H. Yanjun, Method for detecting melamine by aid of nucleic acid aptamers and gold nano-particles, CN104730016A (2015).
[50] Z. Pei, X. Haibo, Z. Shenshan, W. Yuangen, Z. Xuejia, Melamine detection method using a nucleic acid aptamer modified nano-gold particle, CN103217415A (2013).
[51] http://www.who.int/csr/media/faq/QAmelamine/
[52] Y. Huanghao, S. Liang, Z. Xiaolong, T. Zhongqun, Method for detecting synthetic pigments in food by utilizing raman enhanced spectrum, CN103335995 (2015).
[53] L. Xiaoyan, S. Xiaomeng, L. Junjie, L. Min, L. Fuhou, L. Pengfei, Core-shell rhodamine 6G molecularly imprinted solid phase extraction magnetic material, and preparation method and application thereof, CN103433007A (2013).
[54] N.P. Kim, A vessel the bottom absorption a pad food wrapping, KR20060018080A (2006).
[55] K. Nehru, M. Sivakumar, S.C.G.K. Daniel, V. Sureshkumar, Nanoparticle impregnated antimicrobial ice for food preservation, IN03899CH2012A (2014).
[56] C. Quansheng, L. Huanhuan, Z. Jiewen, O. Qin, Nanofiber material for pork preservation, CN103757727A (2014).
[57] Zhang Ming, Nanometer silk fibroin-nano-silver antibacterial composite material, CN103665894A (2014).
[58] M. Cruz, E. Cummins, J. Kerry, M. Morris, Antimicrobial food package, US2015175336A (2015).
[59] J.R. Baker, Jr., S.K. Choi, A. Myc, Bacteria targeting nanoparticles and related methods of use, US2015306238 (2015).
[60] L. Defu, G. Liming, L. Weijie, L. Derong, Z. Wenbo, Nano composite soy protein plastic and preparation method thereof, CN104861662A (2015).
[61] Zhang Ming, Soybean protein isolate-polyaniline antimicrobial composite film, CN103665895A (2014).
[62] E.K. Song, K.S. Jo, J.H. Lee, T.K. Yun, K.R. Lee, M.W. Moom, S.J. Kim, Food container having nano-structured hydrophobic surface and manufacturing method thereof, US2014246429 (2014).
[63] L. Zhaoling, G. Ying, W. Hui, Y. Chongwen, Preparation method of modified ramie nano-cellulose and polylactic acid composite film, CN103992493A (2014).
[64] D. Quintanar Guerrero, M. de la L. Zambrano Zaragoza, A. Alvarez Cardenas, E. Mercado Silva, Composition of solid lipid nanoparticles for the long-term conservation of fruits, vegetables, seeds, cereals and/or fresh foodstuffs using a coating, US2014205722 (2012).
[65] Y. Di, C. Changfeng, W. Bin, Application of *Thamnaconus modestus* fish skin nanoparticle composite edible membrane, CN104844840A (2015).
[66] K. Flemmer, M. Berlin, S. Busoli, Method of providing a package with a barrier and the package thus obtained, US2005208238 (2005).
[67] Y. Zheng, N. Cao, J. Lin, Waterproof and oxygen-insulating sealing film as well as preparation method and application thereof, CN104212053A (2014).
[68] Z. Jinfeng, T. Xiaoli, X. Li, Z. Jiong, Z. Fusheng, Nano-silver/lignin composite antibacterial material and preparation method thereof, CN103749535A (2014).
[69] Y. Zhen, H. Fang, Antibacterial plastic container, CN104118642A (2014).
[70] L. Houbin, L. Qu, W. Heng, S. Jiao, L. Xiao, L. Xinghai, Degradable ppc [chlorinated polypropylene]/SiO_2 nano composite membrane and preparation method thereof, CN103788613A (2014).

INDEX

F

G

H

I

L

M

FSC
www.fsc.org
MIX
Papier | Fördert
gute Waldnutzung
FSC® C083411

Kolibri 360 GmbH
Ferdinand-Jühlke-Straße 7
99095 Erfurt, Deutschland
produktsicherheit@kolibri360.de